The Earth On Trial

The Earth On Trial

Environmental Law on the International Stage

Paul Stanton Kibel

Routledge
New York
London

Published in 1999 by

Routledge
29 West 35th Street
New York, NY 10001

Published in Great Britain by
Routledge
11 New Fetter Lane
London EC4P 4EE

Printed in the United States of America on acid-free paper.
Text Design by Debora Hilu

Library of Congress Cataloging-in-Publication Data

Kibel, Paul Stanton.
The earth on trial : environmental law on the international stage / by Paul Stanton Kibel.
p. cm.
Includes bibliographical references.
ISBN 0-415-92143-0 (hardcover). — ISBN 0-415-91995-9 (pbk.)
1. Environmental law, International. 2. Conservation of natural resouces—Law and legislation I. Title.
K3585.4.K53 1998
341.7'62 — dc21 98-23039
CIP

KMS-

The Rockies may crumble
Gibraltar may tumble . . .

Contents

Publication Credits
First Ink

An expanded version of "City Limits" was published in the *Boston College Environmental Affairs Law Review* (March 1998), under the title "The Urban Nexus: Open Space, Brownfields and Justice."

An earlier version of "Roughshod" was published in *Legal Times* (July 1, 1996), under the title "Brandishing a New Axe: The Salvage Logging Rider and the Constitution," © *Legal Times* 1996. Reprinted with permission of the *Legal Times*, 1730 H Street, NW, Suite 802, Washington, D. C., 20036, 202/247=0682.

An earlier version of "Blaming Wildlife" was published in *In These Times* (April 3, 1995), under the title "Natural Born Killers: Conservatives Attack the Endangered Species Act."

An earlier version of "Words to Choke On" was published in *The San Francisco Recorder* (February 3, 1995), under the title "Debating Greenspeak."

An earlier version of "Ignorance Abroad," co-authored with Court Purdy, was published in *Legal Times* (February 13, 1995), under the title "Exporting Environmental Ignorance," © *Legal Times* 1996. Reprinted with permission of the *Legal Times*, 1730 H Street,

NW, Suite 802, Washington, D. C., 20036, 202/247=0682.

An earlier version of "Axe to the Myth" was published in the *Fordham Environmental Law Journal* (Summer 1995), under the title "Canada's International Forest Protection Obligations: A Case of Promises Forgotten in British Columbia and Alberta."

An earlier version of "Ecology After the USSR" was published in the *Georgetown International Environmental Law Review* (Winter 1994), under the title "Russia's Wild East: Ecological Deterioration and the Rule of Law in Siberia." It is reprinted with the permission of the publisher, Georgetown University, and the *Georgetown International Environmental Law Review*, © 1994.

An earlier version of "United by Poison," co-authored with Armin Rosencranz, was published in *Economic and Political Weekly* (July 2, 1994), under the title "A Blanket Spread Too Thin: Compensation for Bhopal's Victims."

An expanded version of "Refoliating Vietnam" was published in *Environmental Policy & Law* (August/September 1995), under the title "Vietnam—Legal Reform and the Fate of the Forests."

An expanded version of "A Difficult Swim" was published in the *UCLA Journal of Environmental Law & Policy* (Winter 1996), under the title "Justice for the Sea Turtle: Marine Conservation and the Court of International Trade."

Earlier versions of "Trees Falling" were published in *Headwaters Journal* (Winter 1995–96), under the title "The Moveable Feast—Why the Timber Trade is Devouring the World's Forests," and in the *New York University Environmental Law Journal* (December 1996), under the title "Reconstructing the Marketplace: The International Timber Trade and Forest Protection." Reprinted with the permission of the *New York University Environmental Law Journal* and Headwaters Environmental Center, POB 729, Ashland, Oregon 97520, www.headwaters.org.

An earlier version of "The Depths of Europe" was published in *Environs* (May 1996), under the title "Wisdom Across the Atlantic: North America and the European Experience."

Acknowledgments

Many thanks to all the people who encouraged and assisted me in the writing of this book. This includes colleagues who provided critical and insightful review of earlier drafts. This includes my editor at Routledge, Amy Shipper, who helped transform the work from a collection of esoteric essays into a unified text. This includes my friends who did not laugh when I told them I was writing a book that would make environmental policy sexy. This includes Daniel, Kirsten and Chelsea, who provided the technology and workspace to finish the final draft. This includes my parents and my family, who, unlike me, never doubted that this project would someday end up as a book in print. Most importantly, however, this includes my wife Karen, who stuck with me as I followed the strange muse of ecopolicy (one of the lesser-known muses) that led to this book. Her love and patience made all the difference.

Acronym List

B.C.	British Columbia
CERCLA	Comprehensive Environmental Response, Compensation & Liability Act (U.S.)
CIT	Court of International Trade (U.S.)
CITES	Convention on the International Trade in Endangered Species
CSD	Commission on Sustainable Development
EEA	European Environmental Agency
ECJ	European Court of Justice
EPA	Environmental Protection Agency (U.S.)
ESA	Endangered Species Act (U.S.)
EU/EC	European Union/European Community
FAO	Food & Agricultural Organization (United Nations)
FWS	Fish & Wildlife Service (U.S.)
GATT	General Agreement on Tariffs & Trade
GEF	Global Environmental Facility
IMF	International Monetary Fund
IPS	Institute for Policy Studies
ITTA	International Tropical Timber Agreement
MOSTE	Ministry of Science, Technology & Environment (Vietnam)
NAAEC	North American Agreement on Environmental Cooperation
NACEC	North American Commission on Environmental

	Cooperation
NAFTA	North American Free Trade Agreement
NEJAC	National Environmental Justice Advisory Committee (U.S.)
NEPA	National Environmental Policy Act (U.S.)
NMFS	National Marine Fisheries Service (U.S.)
NPDES	National Plan for Environment & Sustainable Development (Vietnam)
NRDC	Natural Resources Defense Council
ODA	Overseas Development Agency (Japan)
OPIC	Overseas Private Investment Corporation (U.S. Agency)
SCCI	State Commission on Corporate Investment (Vietnam)
SCLDF	Sierra Club Legal Defense Fund
SDA	Swedish Development Agency
TED	Turtle Exclusion Device
TDA	Trade and Development Agency (U.S.)
UNEP	United Nations Environment Programme
USAID	United States Agency for International Development
USTR	United States Trade Representative
WWF	World Wildlife Fund
WTO	World Trade Organization

Introduction
Sharp Teeth

Writing, like all things, has its own headwaters, its own sources of origin. To deny these sources is to cut oneself off from the very elements that led one to think and write in the first place. In terms of this book, two particular headwaters are of great importance, for they helped determine both the direction and objectives of the work. To explain where I am headed, I must first reclaim these sources.

In 1987, Russell Jacoby published a book entitled *The Last Intellectuals,* in which he noted, and mourned, the withdrawal of the "public intellectual." Jacoby's central point was not that the modern mind had de-evolved, that it had become somehow less imaginative, less perceptive or less moral. Rather, his point was that the best modern minds had chosen to, or perhaps been forced to, retreat from the public stage. Instead of engaging in debate with society at large, they were instead engaged in debate among themselves.

This internal debate was draining the vitality of the external debate. Increasingly, the public space—where theory and reality are forced into close and often fertile proximity—was being abandoned. A disheartened Jacoby observed that "as intellectuals became academics, they have no need to write in a public prose; they did not, and finally they could not." He warned that this retreat of language was the real danger and the threat, in that "the public relies on a dwindling band of older intellectuals who command the vernacular that is slipping out of the reach of their successors."

Jacoby's message was a condemnation and a lament, but it was also a call to arms. It was a challenge to look beyond our professional peers, beyond the page of scholarship, and to confront the larger public. In terms of the use of language, and the aims of writing, it proposed an important change of focus; to move from the technical to the essential, to employ a strategy of words that would widen and deepen the circle of debate.

This book represents an effort to move in the direction that Jacoby outlined. The debate over the relationship between law and the environment has become increasingly inward looking, with specialists talking more and more to each other. The writings in this book seek to redirect this discussion outward.

If Russell Jacoby is the first headwaters for this book, then the second is Charles Wilkinson. In the field of natural resources law, Wilkinson has played a key role in forging a new language to talk about how society and government interact with the natural environment. Through his writings, he has worked to create a language that rejects legal abstractions to discuss non-abstract phenomenon, yet recognizes the historical and philosophical in even the most legalistic issues.[1]

At a 1991 lecture at Willamette Law School in Oregon, Wilkinson delivered a talk on the Colorado River entitled "Land of Fire and Water." Ostensibly, his topic was western water law. However, his legal discussion included Native American poetry, the geological history of the river canyon, and a survey of the impact of water projects on culture and values. At the close of his lecture, Wilkinson proclaimed: "The language of the law as we now know it is too small to talk about these issues. We need to create a new language for the law, one that is big enough to confront the resource issues that now face us."

Like Jacoby, Wilkinson's message was a condemnation and a lament, but it was also a challenge. It called for environmental and natural resource lawyers to talk plain and to talk deep. Don't say "intensive timber harvesting" when you mean "forest destruction".

Don't say "lawful taking" of animals when you mean "killing." Don't say "resettlement project" when you mean "gunpoint eviction." Don't say "adversely impacted" when you mean "poisoned."

Moreover, Wilkinson urged an open recognition of the moral, the sacred, and the wild. These are the underlying values that prompted the development of environmental law, yet somehow these values found themselves increasingly excluded from the legal vernacular. Wilkinson called for an end to this exclusion.

Taken together, Jacoby and Wilkinson left me with a task: to develop new writing strategies to bring the law-ecology debate into the public space. The writings in this book are an attempt to meet this task.

Therefore, although this book focuses on the law, I did not write this book for lawyers. Although this book focuses on protecting the environment, I did not write this book for environmentalists. The audience I am writing for includes lawyers and environmentalists, but it includes many others.

It includes all persons who understand that the law is fundamentally an expression of public values, and that public values are forged through public debate. It includes all persons who are troubled by the continuing ecological degradation of our world, and by the role our public institutions and private corporations play in this degradation. It includes all persons who believe we have a responsibility to assess the impacts of our actions. It includes all persons who suspect that our future depends not so much on our ability to alter nature to accommodate society, as on our ability to alter society to accommodate nature.

For nature has its own methods of showing us its teeth, of letting us know when we have transgressed limits. The very land, air and water on which we rely begins to turn sterile and toxic. The sum of our transgressions push ecosystems and species beyond the threshold of adaptation, and they begin to die and disappear. This sterility, toxicity and extinction, in turn, degrades not only our natural environment but our economic prospects. It is because of nature's sharp teeth that me must create laws and institutions with sharp teeth of their own. At this point, given our knowledge of the consequences, the aspirational environmental rhetoric of corporations, politicians, and international treaties is not enough. We must work to transform this rhetoric into enforceable, ecologically-sound governance.

One final introductory note on the book. The articles and essays upon which this book is based analyzed issues as they were happening, not after. Most were published as part of ongoing policy research or advocacy projects. These projects were focused on documenting, as well as impacting, environmental and legal developments. As such, there is a strong element of journalism to the text—the pieces are rooted in a specific time and place. And unlike abstract theory or settled history, in journalism there is always risk. Events evolve and issues transmute, so that some aspects of the initial analysis later prove less central and timely, or even incorrect. This was certainly the case with some of the writings in this book.

Such, however, is the risk one takes when one engages with the world as it happens, as opposed the world outside of or after what happens. I accept these risks as a necessary part of, as the author and essayist Terrence Des Pres put it, "writing into the world."[2] These bruises, although not pretty, come with the territory.

To reclaim a deeper and more public language. To place law within the greater context of ecology. To take the events of the present, with all their uncertainty and potential, as a starting point. This book was written with these objectives plainly in mind.

Paul Stanton Kibel
San Francisco, December 1997

Part I.
THE AMERICAN BACKYARD

Since the end of World War II, the United States has established itself as the moral spokesperson for the larger international community. This has been particularly true in the environmental arena, where the United States has generally been the loudest voice condemning the insufficiency of other nations', particularly developing nations', policies.

This role of world environmental monitor has often distracted the United States from taking a closer look at what is happening within its own borders. It has enabled us to operate on the somewhat specious assumption that, in the environmental field, we have answers and other nations have problems. This role of world environmental monitor has also enabled Americans to avoid assuming responsibility for the role our government's policies, and our corporations' activities, play in the environmental degradation of other nations.

Given this tendency, the American backyard is a logical starting point for a foray into the international context. It highlights that the debates about the global environment and world trade are extensions of natural resource and economic conflicts within our own county. It reminds us that there are many critical environmental problems in the United States for which we have not found answers, and for which we may have more to learn than to teach.

Chapter 1

City Limits
Urban Ecology and Economic Justice

America's urban cores, particularly in our older cities, are in a troubled and declining condition. There is clear evidence of the situation, of the general trends that are at work. For several decades the U.S. population has been moving from urban centers to suburban locations, and the number of citizens living in the suburbs now exceeds the number of citizens living in the cities. Open space surrounding urban centers is rapidly being converted to residential and commercial use, while large tracts of urban housing and commercial property are now vacant, polluted or both. The gap in per capita income between urban residents and suburban residents is growing ever wider, and the crime and unemployment rates in urban areas are growing higher.[1] Minority populations in declining urban cores have become more geographically isolated, creating a situation of de facto segregation. As the city tax base declines, municipal governments have less resources to support education, police and other essential services.

These observations are not offered in support of any argument. They are simply a summary of a broad economic, environmental and racial phenomenom that most of us have witnessed with our own eyes and experienced in our own lives. In its most condensed form, this phenomenon is as follows: Jobs and people are moving out of urban centers into formerly pristine surrounding areas, leaving behind polluted vacant lots and unemployed minority populations.

Although it is not too difficult a task to describe the reality of urban decline, it is another task altogether to identify and isolate the underlying trigger of this decline. Many different culprits have been proposed, including racism, capitalism, industrialism decline, technology, drugs, the media, the automobile, the police, the public school system, too much government regulation, and too little government regulation. Is one entity or issue ultimately responsible for why our cities are now subject to such powerful and destructive economic, environmental and racial pressures?

This is an important question, but a question that will not be answered here. Regardless of whether there was initially an underlying trigger, we have reached a point where the various components of urban decline are now feeding on and reinforcing each other. They are all interconnected contributors to the downward spiral that has left our urban cores in their current condition. Therefore, instead of arguing for or against a particular underlying cause, our task will be to focus on the relation between certain critical components of the cycle of urban decline. More specifically, three particular components will be assessed: suburban sprawl and open space loss, hazardous waste liability, and health conditions and economic welfare of communities living in the urban core. Although this analysis will draw extensively on the experience in the San Francisco Bay Area, the issues discussed are not specific to any particular city; they are affecting virtually every major U.S. metropolitan area.

To be certain, open space loss, toxic contamination and economic inequity are not the only components of the urban decline cycle. However, they are three areas in which existing law, especially in terms of land-use zoning and environmental liability, has played a crucial role. They are therefore also areas where legal reform can potentially play a crucial role in reversing the pattern of urban decline. An historical and policy framework is needed, however, to effectively evalutate such reforms.

Open Space and the City

For the past half century, there has been one dominant model for metropolitan growth in the United States. It has been described as "unlimited suburban sprawl"[2] or "low density discontinuous development."[3] The basic component of this metropolitan paradigm has been the conversion of wilderness and farmland, commonly called open space, to commercial and residential use. In this conversion scenario, the emphasis has been on the development of shopping centers and business/industrial parks for commercial use, and detached, single family homes with yards for residential use.

Before turning to the present day economic and environmental consequences of this development pattern, its origins must first be revisited. In the modern context, the terms "city" and "suburb" have taken on very strong political and cultural meanings. As Zignew Rybczynski, an urban historian at the University of Pennsylvania, explained in his 1995 book *City Life*, the two terms "are often only polemical categories: depending on your point of view, either bad (dangerous, polluted, concrete) cities and good (safe, healthy, green) suburbs, or good (diverse, dense, stimulating) cities and bad (homogeneous, sprawling, dull) suburbs."[4] Beneath these polarized meanings, however, there is a great deal of historical and ideological undergrowth. We must examine this undergrowth to make sense of where we are today, to place the debate over open space conversion and the exploding metropolis in a broader context.

Although the conversion of open space to commercial and residential use tends to be thought of as a recent trend, it is in many ways a continuation of a deeply ingrained American tradition—the frontier. For hundreds of years, the American experience involved the push westward across the continent, of clearing wilderness and breaking the land. The American frontier provided an outlet for those who were dissatisfied with their economic or social prospects in a given location; they could vote with their feet, by moving west to a less congested, less socially stratified, or less expensive region of the country.

The outlet of the frontier played a critical role in shaping the U.S. economy and American society. It meant that the upward mobility of the lower classes need not come at the direct expense of the more established upper class, because lower classes could seek their fortunes elsewhere rather than directly challenging those already wealthy. It meant that Americans were less tied to geographic place, and therefore when confronted with regional problems, they were more likely to move than seek place-specific solutions. The national experience with the western frontier helped establish some of the values and patterns that would later lead to suburban sprawl and urban decay.

The forces that would contribute to the geographic decentralization of urban areas were identified early on. In 1900, H.G. Wells published a prophetic essay entitled *The Probable Diffusion of Great Cities*. In this essay, Wells forecast that urban regions would become so vast that the very concept of the city would become "as obsolete as the mailcoach." From Wells's perspective, this diffusion was not altogether negative. It offered people the possibility of healthier and less congested lives, and of an alternative to the disease and filth that often characterized turn-of-the-century industrial cities.

As Wells's 1900 essay suggests, initially the concept of suburbs and suburbanization did not carry with it the cultural and environmental stigma that it carries today. The first generation of suburbs in the United States, which included such areas as Philadelphia's Chestnut Hill, Chicago's Lake Forest and Cleveland's Shaker

Heights, bore little resemblance to many of the suburbs of today.[5] Unlike the sprawl of contemporary suburbia, the first generation of suburbs in America were equated with innovative land-use planning, high-quality architecture, pedestrian access, and good suburban-urban public transportation (usually by train). In fact, it was the success of these early "garden suburbs" that created the market for, and the allure of, suburbanization. Prior to Chestnut Hill, Lake Forest and Shaker Heights, the American dream, at least residentially speaking, focused mostly on the city, the farm or, perhaps if one were rich enough, the country estate. The garden suburbs of the early twentieth century moved the suburban ideal towards the center of the American identity.

The tragedy is that the very characteristics that drew people to the first generation of suburbs began to disappear as their popularity increased and more people moved out of the city. Suburban developments began to fill in the open space, degrading scenic views and reducing undeveloped natural areas and farmland. Priorities such as land-use planning, quality architecture, and good suburban-urban public transportation were sacrificed to meet the growing demand for low-cost suburban housing. The garden suburb gave way to the subdivision, the shopping mall and the freeway, and suburbanization began to take on a new and more ominous meaning. Although initially envisioned as a means to escape the congestion of the city, the suburbanization process eventually created is own brand of overgrowth—decentralized congestion. As Lewis Mumford observed in 1961, "The ultimate effect of the suburban escape in our time is, ironically, a low-grade uniform environment in which escape is impossible."[6] Mumford continued, "A universal suburb is almost as much a nightmare, humanely speaking, as a universal megopolis; yet it is toward this proliferating nonentity that our present random or misdirected urban growth has been steadily tending."[7]

Lewis Mumford's critique of suburbanization was based largely on aesthetic and cultural grounds, on the dull and prefabricated landscape contemporary suburbia tends to create. His critique is closely related to this article's central point. There are identifiable reasons why cities have traditionally served as important cultural centers. The reasons include the face-to-face interaction of people from different economic classes and ethnic backgrounds, the architectural and historical heritage of neighborhoods and city centers, and the maintenance of parks, commons and other public spaces. Land-use zoning, open space preservation, environmental liability and justice—the issues addressed here—provide the legal framework that helps determine whether these urban amenities will endure or decline.

With this historical context in place, we can now turn to the modern consequences of, and responses to, the exploding metropolis. Environmentally and economically, the impact of suburbanization has been profound. Environmentally, commercial and residential development has now pushed deep into natural canyon, coastal and woodland ecosystems, with a corresponding loss of habitat for wildlife and public recreation areas for people.[8] The conversion of farmland to subdivisions and industrial uses has destroyed beautiful landscapes and displaced rural communities. The lack of adequate public transportation, the reliance on automobiles, and the increasing distance of commutes has also led to severe air pollution in many metropolitan areas.

Economically, the impacts of suburbanization have been mixed. For the automobile and construction industries, and for the treasuries of many suburban municipal governments, it has been a boon. For city centers, however, it has created many problems. As businesses and residents have left for the suburbs, cities have seen a decline in tax revenues and municipal services, and a rise in unemployment and crime.

Although this shift in fortunes between cities and suburbs initially seemed justified on market grounds, it has become increasingly clear that this shift has created new economic problems. As urban unemployment rises, the rest of society, including those in the suburbs, are required to fund state and federal welfare assistance programs. As air quality declines, as open space vanishes, and as malls and subdivisions come to dominate the landscape, the region becomes less desirable as compared with other regions. As a result, home buyers and businesses choose to relocate to these other regions. Thus, over time, the economic welfare of the entire metropolitan area begins to suffer: The problems of the city begin to pull the suburban economy down with it.

The economic, environmental and political unsustainability of suburban efforts to disengage from urban cores has been recognized not only by open space and urban poor advocates, but by the business community as well. In 1995, Bank of America, the largest bank in California and one of the largest banks in the United States, co-published a major report entitled *Beyond Sprawl*. In this report, Bank of America concluded that "unchecked sprawl has shifted from an engine of California's growth to a force that now threatens to inhibit growth and degrade the quality of our life," and that "allowing sprawl may be politically expedient in the short run, but in the long run will create social, environmental and political problems that we may not be able to solve."[9]

Similarly, in 1991 the Bay Area Council, a policy organization representing major employers and businesses in the San Francisco region, published a report on growth management.[10] In its report, the Bay Area Council argued that current growth patterns would lead to "economic and environmental decay" in the area, and that new strategies were needed to protect open space. The report even went so far as to suggest the creation of a Bay Area Greenbelt, a ring of undeveloped open space surrounding the entire metropolitan area.

In response to the problems created by sprawl, local governments and communities have developed strategies to control suburban growth. Three of the most widely used strategies for controlling sprawl are slow-growth initiatives, residential lot requirements, and private land trusts. Slow-growth initiatives place an absolute percentage limit, or even an absolute moratoria, on the amount of new residential units that can be built in a given time period. Residential lot requirements establish rules regarding the size or type of new residential construction, such as only single family homes with a minimum amount of acreage. Private land trusts enable local citizens to collectively purchase open space or farmland, and thereby prevent such properties from being converted to commercial or residential use.

Slow-growth initiatives, residential lot requirements, and private land trusts have helped individual communities block the development of new, less upscale, housing. However, they have not addressed the problems that are prompting urban flight, nor have they prevented sprawl from leapfrogging over regulated slow-growth areas to other undeveloped and less regulated areas.[11] Moreover, in many instances, local anti-sprawl measures were based more on a concern for property values than for open space preservation.[12] The environment was often only a pretense for the rich to exclude the poor and middle class from certain neighborhoods.[13] In such situations, the economic inequities initially created by sprawl were only intensified by local efforts to stop it.

The U.S. experience with suburbanization and open space conversion has taught environmentalists, urban poor advocates, policy makers and the business community an important lesson. Suburbanization may provide select individuals and companies with a short-term escape from the problems of urban decline, but it does not provide society with a long-term policy solution. In the long run, we cannopt simply move away from the problems affecting the city, because these problems eventually impact us all. Economically and environmentally, the paradigm of the exploding metropolis, of suburbs geographically and politically segregating themselves from the city, cannot be sustained.

Brownfields Under Superfund

The abandoned, deteriorating property has become a dominant image of our cities. It has come to represent the ghost town quality, the so-called blight, of so many of our urban areas. The vision of the vacant urban lot embodies most of the elements commonly associated with the decline of our cities: pollution and garbage, unemployment, poverty, racial isolation, crime, drugs, declining public services, and architectural eyesores.

As discussed earlier, the causes of the vacant urban lot, and of urban decline in general, cannot be readily reduced to a single issue. While there may have been an initial cause or trigger, we have now reached a point where several factors are reinforcing the process of abandonment, decay and disinvestment. One of the most significant factors in this process is the liability associated with properties that are perceived to be, or are in fact, contaminated with hazardous materials.

Liability for the cleanup of contaminated property is established primarily under federal and state environmental laws. The most far-reaching of these laws is the 1980 federal Comprehensive Environmental Response, Compensation and Liability Act (CERCLA).[14] CERCLA is often referred to as Superfund, after the revolving cleanup trust fund established under the law. Most of the state hazardous waste cleanup laws were based largely on the federal Superfund model. Therefore, by examining Superfund we can observe how environmental liability laws in general are affecting the use or abandonment of urban properties.

CERCLA's core objective is to identify parties responsible for contaminating property, and to then require these parties to directly pay, or reimburse the government (usually the Environmental Protection Agency), for the costs of environmental remediation. Under the law, persons who are subject to remediation liability are referred to as potentially responsible parties, or PRPs. On its face, CERCLA appears as a workable and appropriate piece of legislation; a straightforward law based on the polluter pays principle, which holds that the burden of cleanup should fall on the shoulders of those who pollute. In practice, however, CERCLA has proven difficult and somewhat dysfunctional.

CERCLA's troubles can be traced in part to the expansive interpretations of liability adopted by EPA and the courts. These expansive interpretations resulted in the following liability rules: (1) *strict liability*, in which intent or negligence were not required to impose remediation liability; (2) *joint and several liability*, in which a party who contributed to a small portion of the pollution could be responsible for the entire cost of remediation; (3) *lender liability*, in which banks and lending institutions that influenced the management decisions of property owners could be subject to cleanup liability; (4) *retroactive liability*, in which a party could be subject to cleanup liability notwithstanding that its hazardous waste disposal practices were legal at the time the disposal occurred; and (5) *open-ended liability*, in which a party remained uncertain when remediation was completed, or what cleanup standards would satisfy its remediation responsibilities.

Although CERCLA's expansive liability rules were intended to facilitate comprehensive and speedy cleanup of contaminated sites, often this was not result. Frequently, the liability was so extensive that parties found it cheaper to litigate for years rather than to pay for remediation. Frequently, the specter of lender liability meant that banks would refuse to foreclose on loans and properties would be abandoned. Frequently, investors and banks would refuse to redevelop contaminated property, or even property that might be contaminated, for fear of becoming a liable party. Frequently, landowners would avoid undertaking a preliminary environmental assessment of their property, because such an assessment could unearth information that might trigger cleanup liability.

Under the above liability scenario, environmental lawyers and remediation consultants were making substantial profits. Despite the enormous activity surrounding CERCLA's implementation, however, there was often a disturbing lack of activity on the actual remediation front.[15] Lawyers and consultants were hired to help determine CERCLA remediation liability, but much of their work never translated into tangible cleanup of contaminated properties.

The subject matter of CERCLA, the polluted sites, generally remained just that—polluted sites. Especially in former industrial urban areas, the American landscape remained littered with abandoned, contaminated properties. Although CERCLA environmental liability was certainly not the only factor contributing to this situation, it nonetheless helped deepen the post-industrial economic decline in many city neighborhoods. From an investment and business standpoint, these abandoned properties, or brownfields, became untouchables.[16]

Abandoned brownfields tended to drag surrounding properties and communities down with them, thereby reinforcing the decline cycle. As discussed earlier in this article, the increase in untouchable brownfields also encouraged suburban sprawl and the destruction of open space. This pattern of metropolitan expansion only further diminished many cities' economic resources and political power.

The point here is not to blame CERCLA for the woes of post-industrial urban America. Rather, the point is simply to demonstrate the particular role that environmental liability rules played in diverting investment and economic development away from our cities.

In response to the economic and environmental problems relating to PRP liability rules, there have been some attempts to reform CERCLA. The first significant attempt to reform CERCLA was the Superfund Amendments and Reauthorization Act (SARA) of 1986.[17] Among other things, SARA sought to establish a viable "innocent landowner defense" for parties who purchased property after contamination occurred. Under SARA's provisions, a purchaser would not be liable for remediation costs if the party could demonstrate that it "did not know nor had reason to know" of the hazardous waste contamination when the party acquired the property. The objective of this language was to provide the prospective purchaser with sufficient protection, or immunity, so that polluted properties could be redeveloped.

Due to inconsistent interpretations of the innocent landowner defense, however, SARA did not achieve this goal. More specifically, EPA and the courts did not clearly establish what a prospective purchaser must do, in terms of environmental investigation, to demonstrate that the party "did not know nor had reason to know" of existing contamination. In the absence of such specific criteria, SARA's protections could not be relied upon. As one commentator explained, in practice CERCLA's innocent landowner defense turned out to be more of a mirage than an oasis.[18] As a result, acquisition and redevelopment of polluted properties did not occur, and the untouchables remained largely untouched.

The second major wave of CERCLA reform, EPA's Brownfields Action Agenda (EPA Agenda), began in late 1995 near the end of President Clinton's first term.[19] The EPA Agenda emerged from the ashes of the proposed 1994 Superfund Reform Act, a Clinton-sponsored bill which congress did not pass. In the absence of strong congressional action, the focus of CERCLA reform shifted to the administrative arena. What could not be achieved through broad-based legislation would now be attempted through a package of agency policies and operating procedures.

Prior to the EPA Agenda, the term brownfield generally held a negative meaning, both environmentally and investment-wise. It referred to former industrial properties that were now unused due to uncertainty over environmental remediaton liability. EPA's program sought to transform this meaning, to change the language of brownfields from talk of obstacles to talk of opportunity. An April 1996 report issued by EPA reflects this shift: "The Brownfields Action Agenda will help reverse the spiral or unadressed contamination, declining property values and increased unemployment often found in inner city industrial areas."[20] As such, the EPA Agenda suggested that the brownfields issue was not just about limiting the liability of banks and real estate developers; it was also about providing inner-city residents with a strategy to improve the economy and environmental health of their communities.

The EPA Agenda called for several changes in agency policy and operating procedures. These changes included, among other things: (1) removing thousands of properties from the national tracking list of contaminated sites;[21] (2) *prospective purchaser agreements*, in which EPA agreed not to sue new owners for cleanup of contamination that occurred prior to purchase; (3) *land use-restrictions*, in which new owners agreed to limit future use to commercial and industrial purposes, in exchange for EPA's release of cleanup liability; (4) *national and regional brownfields pilots*, in which EPA provided grants to states and local governments to help cleanup and redevelopment contaminated properties; and (5) *Community Reinvestment Act (CRA) credits*, in which banks could fulfill CRA's local-lending obligations by providing loans for brownfields cleanup and redevelopment.

In addition to the EPA Agenda, recently federal legislation was passed that could provide further liability protection for banks and other lending institutions. The 1996 Asset Conservation, Lender Liability and Deposit Insurance Protection Act , creates a new "lender exemption" under CERCLA. This exemption permits banks to take certain specified actions without triggering Superfund liability. These exempted actions include foreclosure, resale and leasing of the premises. Although there still remain many actions that could trigger Superfund liability, especially lender actions that might influence how a landowner manages environmental problems on a given site, the federal legislation does provide greater clarity and certainty. At least in regard to the actions specifically exempt, banks and other lending institutions should be better able to determine their liability.

As discussed earlier, CERCLA is not the only law that creates liability for the cleanup of contaminated properties. There are laws in virtually every state that establish CERCLA-type liability schemes for environmental remediation. The policy debates around brownfields reclamation have therefore focused not only on CERCLA, but on state hazardous waste laws as well.

Who Is Reclaimed?

As the previous sections on open space conversion and brownfields reveal, metropolitan land-use and hazardous waste remediation are closely linked to the fate of the urban poor. Because the urban poor often tend to be people of color, these issues also raise difficult questions of equity and justice. How do the location of contaminated sites, and the rules governing environmental liability, impact the economic and health conditions in communities of color? Do the negative economic and environmental consequences of open space conversion affect all ethnic groups equally? Will brownfields reclamation provide tangible benefits, in terms of economic development or environmental quality, for the communities where brownfields are located, or will reclamation mostly benefit investors from outside the community?

The questions presented above all fall within the larger policy issue of what is now generally called "environmental justice." The environmental justice movement is based on the growing recognition that poor communities and minority populations are subject to disproportionately high health and environmental risks. Government policies that have either encouraged or ignored this disproportionate allocation of risks have been justifiably classified as examples of "environmental racism." The goal of the environmental justice movement is to empower disadvantaged communities, and to educate and pressure government agencies, to ensure that environmental protection policies benefit all citizens, not just the white and the rich.

From both a racial and an environmental standpoint, environmental justice is a significant and long-overdue development. The movement represents the convergence of two agendas that traditionally had little interest in or understanding of each other—civil rights and environmental protection. More specifically, it forced the environmental movement to confront some of the racist and class-driven aspects of its political platform. Environmentalists had come to consider environmental protection as something distinct from, or something above, the struggle for justice and equity. By demonstrating that levels of environmental protection were closely related to citizens' race and wealth, environmental justice advocates laid bare the falsity of this position.

By the time EPA began developing its Brownfields Action Agenda, the environmental justice movement was already in high gear. For several years, disadvantaged communities had begun to organize around health and environmental issues, and had managed to force changes in government and corporate policy. Several successful environmental justice law suits and administrative challenges had been filed. Additionally, President Clinton took two actions that helped raise the political profile of the movement. First, in 1993 the National Environmental Justice Advisory Council (NEJAC) was established to provide independent advice, consultations and recommendations to the EPA Administrator on environmental justice matters. Second, in 1994 President Clinton issued Executive Order 12898, which called for federal agencies to take actions to address environmental justice in minority and low income populations.

Given these developments in the area of environmental justice, the push for brownfields reclamation was met with both anticipation and skepticism. On the one hand, brownfields reclamation provided an opportunity to cleanup and improve economic and environmental conditions in many poor and minority neighborhoods. On the other hand, brownfields reclamation also called for less stringent cleanup standards and shielding banks and investors from remediation liability. Furthermore, there were no guarantees that the new jobs made possible by reclamation would go to the people who lived in the communities where brownfields were located. Thus, it was possible that brownfields reclamation could lead to a continuation or worsening of health and economic conditions in poor and minority neighborhoods.

Skepticism about brownfields reclamation was based on more than environmental justice concerns. It was based on previous negative experiences with urban renewal policies. During the 1960s, state and federal governments implemented many programs aimed at improving housing and economic development in inner cities. These programs failed for several reasons. The widespread development of housing projects was thought to have isolated and stigmatized poor minority populations, and to have led to increased crime and segragation. The renovation of older neighborhoods often resulted in gentrification, in which neighborhood residents were priced out of their own communities. The economic development programs were often focused on businesses that did not hire from the community. Thus, the jobs that were created often did not benefit those who lived in the neighborhood. From the perspective of many inner city residents in the areas targeted for redevelopment, urban renewal essentially meant removal of poor and minority people.[22]

Many suspected that the 1990s brownfields agenda would be a repeat of the 1960s urban renewal experience. These concerns were expressed poignantly by Olin Webb, a construction engineer and long-time resident of the Bayview-Hunters Point neighborhood in San Francisco. Bayview-Hunters Point contains numerous contaminated and abandoned properties, and a majority of its residents are minorities. The neighborhood has therefore been a focal point for government and private sector brownfield initiatives in the San Francisco Bay Area. Many of these initiatives have been portrayed by government and investors as community redevelopment projects. Mr. Webb, however, views these initiatives as part of a longer and more disturbing pattern: "As far as I'm concerned, a brownfield is just a Superfund site. African Americans bore the brunt of the poison and pollution when they were Superfund sites, but now they are not going to be a part of cleanup and redevelopment. From my neighborhood's perspective, brownfields redevelopment means that African Americans are being passed over and moved out."[23]

As discussed in the previous section on CERCLA reform efforts, the EPA began developing its Brownfields Action Agenda in early 1995. By this time, the environmental justice movement had become a powerful political force, and President Clinton had recently issued his 1994 Executive Order on Environmental Justice. Thus, at least at the level of government policy, environmental justice and brownfields reclamation became major political priorities at a similar point in time.

The concurrent political ascendance of environmental justice and brownfield issues forced the Clinton Administration to develop new strategies to handle this emerging policy nexus. In terms of a best case scenario, they were looking for ways to stitch the two movements together—to integrate equity, environmental cleanup, and economic revitalization into one coherent and mutually-reinforcing policy agenda. In terms of damage control, they wanted to avoid a situation where the environmental justice and brownfields agendas were in visible contradiction, mutually undermining each other.

The Clinton Administration's first significant effort to integrate environmental justice and brownfields policies took place in the context of the National Environmental Justice Advisory Committee. In 1995, NEJAC and EPA co-sponsored a series of public dialogues on brownfields and urban revitalization. The dialogues were held in five cities (Boston, Philadelphia, Detroit, Oakland and Atlanta) and focused on EPA's plans to adopt new brownfields cleanup and redevelopment policies. NEJAC's public dialogues involved persons from varied backgrounds and with varied objectives. Among those who participated were persons from community groups, government agencies, religious groups, unions, universities, banks and philanthropies.

Although the NEJAC dialogues revealed that there was broad interest in the issue of brownfields, they also revealed the profound gulf of both objectives and language that existed between different stakeholders. The word "redevelopment" was being used by all the participants in the NEJAC dialogues, but the term clearly meant different things to different people. For the real estate investors and banks, redevelopment meant removing the liability risks associated with property transactions at sites where there were toxic contamination concerns. For environmental justice advocates, redevelopment meant ensuring that health conditions and the economic self-reliance of poor, inner city residents were improved, not worsened, by brownfields reclamation.

The divisions that emerged at the 1995 NEJAC dialogues have continued to define the evolution of the brownfields issue. In the San Francisco Bay Area, for instance, many of the local participants in the Oakland NEJAC dialogue went on to form the San Francisco Bay Area Regional Brownfields Working Group (SF Brownfields Working Group). Although the group includes members from the lending, business and regulatory communities, the main focus of the group's work is to promote environmental justice in the context of the brownfields issue, to strengthen community-leadership and participation in efforts to redevelop contaminated properties.

To help advance these environmental justice goals, in May of 1997 the SF Brownfields Working Group organized a brownfields workshop entitled "Community Development & Environmental Restoration." Unlike the 1995 NEJAC dialogues, the SF Brownfields Working Group workshop was not designed to help EPA formulate new hazardous waste cleanup policies. Rather, the goal of the 1997 workshop was to educate community leaders on existing government policies and lending/financing options in the brownfields area. It provided information on how local non-profits and small businesses can take the lead, and leverage resources, to clean up sites and put them back into productive use. As such, the focus of the workshop was on helping neighborhoods to become the initiators, rather than the victims, of brownfields reclamation.

At the same time as groups like the SF Brownfields Working Group are pushing ahead on the environmental justice front, other stakeholders are seeking to frame the brownfields issues in terms of pure investment opportunities. For instance, in March of 1997, a new monthly national magazine, *Brownfield News*, was launched in Chicago. The magazine proclaims itself to be "The Source of the Distressed Property Market," and contains articles on industrial real estate forecasts, investor insurance coverage, strategies to reduce expenditures on environmental cleanup, and new legislative proposals to reduce investor and lender liability. In the pages of *Brownfields News*, one is not likely to find discussion of economic equity, public participation or environmental racism. These issues simply fall outside the investment scope of the publication.

The point here is not the portray the SF Brownfields Working Group and *Brownfield News* as two opposite ends on a spectrum of good and evil. Clearly, environmental justice advocates need to access and leverage private capital to achieve their community empowerment goals. Local non-profits and government agencies can take the lead in defining how neighborhood redevelopment should proceed, but only the private sector can provide the financial resources to make these plans work. Given that the private sector will be the ultimate engine of brownfields reclamation, much of the information presented in *Brownfield News* could be used to further the environmental justice agenda. It could be viewed as a tool for helping communities take control of their economic and environmental future.

Despite the potential confluence of interests, however, environmental justice advocates remain wary of the growing role of the lending and investment communities in brownfields redevelopment. As with urban renewal in the 1960s, there is concern that the brownfields issue is being economically and politically highjacked by interests that have no connection with, or true concern about, the communities they claim to be helping. In the language of investors and lenders, struggling communities and poisoned citizens can be readily reduced to the term "distressed property market," a market in which profit alone becomes the governing redevelopment principle.

In the brownfields debate, environmental justice advocates have posed a critical question: Can there be a commitment to urban neighborhoods, economic equity and public health when remediation policy and investment are driven by profit alone? The answer to this question will impact citizens and communities across the nation.

Metropolitan Vantage Point

The origins of suburban sprawl, toxic contamination and inner city decline are complex. Given this complexity, there are no simple policy solutions to these problems. The scope and interrelatedness of the issues do not lend themselves to tidy, reductionist answers.

While there may not be simple solutions, there are nonetheless specific and important policy steps that can be taken to improve the situation. Particularly in the areas of metropolitan land governance and the remediation regulatory framework, there are policy options that can and should be pursued. These options are discussed below.

In the area of metropolitan land governance, there needs to be a recognition that our municipal governments often lack the legal capacity to deal with the problems facing our cities. Jurisdiction over land regulation generally resides at the county level, yet the problems of open space loss and inner city disinvestment frequently operate on a larger metropolitan scale.[24] So long as land-use planning, property taxes and municipal services are handled by county governments, different counties will lack either the means or the incentive to deal with metropolitan wide land-use problems.

Illustrations of the inadequacy of current metropolitan governance are easy enough to find. A county that chooses to protect open space cannot generally prevent a neighboring county from encouraging sprawl, and adding to traffic and air pollution. Inner city counties containing large numbers of contaminated properties cannot require that surrounding suburban counties help fund remediation. In many metropolitan areas, there is no way to ensure that affordable housing is available to middle and lower income residents, because each county is seeking to upgrade its tax base.

The inadequacy of metropolitan governance is particularly acute in the land-use area. As Joe Bodovitz of California Environmental Trust, a non-governmental organization focused on growth-management issues, observed:

> "Sustainable environmental planning comes down to three basic elements: land, air and water. In the Bay Area, the Regional Water Quality Control Board has the region-wide institutional capacity to deal with water quality and the nine-county Bay Area Air Quality Management District has the institutional capacity to deal with air quality. The problem is that there is no region-wide institution with the capacity to adequately deal with land, and without the land element, the environmental quality of the Bay Area cannot be preserved."[25]

In 1991, state legislation was introduced in California that would have helped establish the foundation for meaningful metropolitan governance in the San Francisco Bay Area. The proposed legislation called for the consolidation of several existing regional institutions and agencies into one governmental entity called the Regional Commission. Despite the support of the environmental, minority rights and affordable housing advocates, the bill creating the Regional Commission was rejected by the California legislature. This rejection was due in large part to two factors. First, some existing agencies proved unwilling to transfer authority or funding to the new Regional Commission. Second, many less populated Bay Area suburban communities were convinced that the Regional Commissionís agenda would be dominated by urban interests.

Although recent efforts to strengthen metropolitan-wide governance did not succeed in the San Francisco Bay Area, other cities have had better luck. In 1978, for instance, Portland, Oregon voters approved the creation of a new multi-county agency, the Metropolitan Services Agency, with significant land-use authority. The Metropolitan Services Agency, or "Tri-Met," whose councilors are elected from the city's three counties, has jurisdiction over development, housing and open space preservation for the entire Portland metropolitan area. Portland's multi-county agency has been credited with preventing the sprawl, traffic congestion and affordable housing shortages that have plagued many other cities.

While a Tri-Met-type agency may not be the appropriate solution for all cities, Portland has at least provided an important model for metropolitan governance. The citizens of Portland have demonstrated that it is indeed politically possible to create metropolitan institutions that operate at the same scale as the land-use problems confronting our cities.

In the area of brownfields remediation policy, the critical task will be to place environmental and economic justice issues at the center of the redevelopment process. Through the federal EPA Brownfields Action Agenda, and similar state environmental reforms, the liability framework for contaminated properties is beginning to change. State and federal laws and regulations increasingly offer enhanced protection to investors who are willing to purchase sites with real or perceived hazardous waste problems.

While these investors will likely help put brownfields back into economic use, it remains unclear what impact this redevelopment will have on inner-city communities and the environment.

Governments can play an important role is shaping the redevelopment process. Most significantly, governments can provide a regulatory framework that will point the private sector, the underlying engine of brownfields redevelopment, in a more environmentally progressive and equitable direction. They can refuse to accept lower cleanup and health standards for properties located in poor, inner city neighborhoods. They can develop more powerful tax incentives, along the lines of the federal Community Reinvestment Act, to ensure that brownfield redevelopment loans from private banks are made to businesses from within distressed neighborhoods. They can adopt policies that link cleanup liability protections to whether the proposed redevelopment project will have tangible health and economic benefits to the local community.

One possible model for integrating remediation reform with environmental justice is the federal Small Business Administration.[26] The Small Business Administration establishes a program which encourages federal agencies to favor small business enterprises in the awarding of government contracts, so long as these enterprises possess the capacity and expertise to fulfill the contracts. The program recognizes that: (1) smaller enterprises, because of economies of scale and vertical integration, are often underbid by larger national or international companies; and (2) there are valid policy reasons for providing some degree of protection for these smaller enterprises, which are often owned by and employ workers from the local community. The Small Business Administration provides a means to protect and promote these neighborhood, community focused, businesses.

EPA and other state environmental agencies could establish liability release programs that operate similarly to the Small Business Administration. The decision of whether to release a private party from future cleanup liability could be based, in part, on whether the private party is a community enterprise. EPA and other state environmental agencies could establish a policy wherein community enterprises were expressly favored in the granting of liability releases. This type of liability release program would help promote environmental justice goals by helping ensure that community enterprises participate in the economic benefits of brownfields reclamation.

Another possible model for integrating remediation reform with environmental justice are the Restoration Advisory Boards, created to help deal with environmental cleanup issues relating to military base closures.[27] These boards are charged with helping develop and monitor the remediation process.[28] In theory, their job is to ensure that policies treat military bases as integrated communities rather than a collection of discrete and independent properties. As such, they promote solutions that are responsive to the broader community impacts of base closures and environmental cleanup. EPA and other state environmental agencies could establish community-based boards along these lines, which would then help guide neighborhood remediation policy and objectives. These boards would help ensure that government decisions regarding cleanup standards and liability are dealt with from a community-based, rather than a parcel-by-parcel, perspective.

Through the 1993 Executive Order and the 1995 NEJAC public dialogues, the Clinton Administration has taken several bold symbolic steps in the area of environmental justice. Now what is needed are policies to translate this rhetoric into political and economic reality, so that brownfields reclamation can contribute to the larger reclamation of America's troubled cities. The Community Reinvestment Act, the Small Business Administration and Restoration Advisory Boards provide a starting point for developing and implementing such policies.

The City Frontier

For centuries, the U.S. frontier was about breaking the land, of pushing the geographic edges of development continuously outward. As suburban sprawl, urban decay and environmental pollution have made plain, however, that frontier has reached its end. Ecologically, economically and politically, the paradigm of uncontrolled and continuous outward land development cannot be sustained.

The frontier before us now is about forging new relationships among our cities, farmlands and wildlands. It is about constructing policies and economies that promote the health and livelihood of all our citizens, not just the privileged. The effort to reconcile open space, brownfields and justice issues is on the leading edge of this new frontier. The success or failure of this effort will impact not only the fate of our cities, but the fate of our ecology and economy as well.

Chapter 2

Roughshod

Northwest Forests and the Constitution

The salvage logging rider adopted by Congress in 1995 has been widely criticized by the environmental community. This rider aims to insulate salvage logging timber sales from forestry and wildlife laws through a congressional declaration that such sales were "deemed to satisfy" existing environmental laws.[1] Because this language was intended to prevent application of environmental laws to salvage logging sales, the provision has become known as the "logging without laws" rider.

The environmental community's critiques of the rider have focused mostly on the political and ecological aspects of the provision. Politically, environmentalists have argued that the rider would reverse the legislative progress that forest protection advocates have made over the last twenty-five years, and that it would jeopardize President Clinton's Option 9 plan for the Pacific Northwest, that was developed in 1993 to help resolve the Spotted Owl controversy. Ecologically, environmentalists have argued that the rider would severely damage forest health, and destroy critical habitat for endangered fish and wildlife.

The political and ecological critiques provide solid grounds on which to repeal the salvage logging rider. There may, however, be an even more powerful and fundamental reason to question and condemn the provision. Under Article III of the U.S. Constitution, the judiciary is given ultimate and final authority to determine legal compliance or violation in specific cases or controversies. By including language in the rider that simply declares a certain category of timber sales satisfy (comply with) legal requirements, Congress may have overstepped its constitutional authority. As such, portions of the "logging without laws" rider may violate the separation of powers doctrine, and therefore be unconstitutional.

The Independence of the Courts

Article III of the U.S. Constitution sets forth the judiciary's power to review and rule on cases or controversies arising under the Constitution or the laws of the United States. The U.S. Supreme Court has held that this constitutional provision prevents the executive or legislative branches from impermissibly intruding upon judiciary's ultimate authority to resolve specific legal cases or controversies. The seminal U.S. Supreme Court decision on the question of when Congress has impinged on the Court's traditional role is *United States v. Klein*.[2]

In *United States v. Klein*, the Court considered the constitutionality of a 1870 legislative provision that sought to remove judicial jurisdiction over cases currently pending before the U.S. Supreme Court. Writing for the Court, Chief Justice Salmon P. Chase concluded that "Congress had inadvertently passed the limit which separates the legislative from the judicial power by actually directing the disposition in a particular case." The *Klein* decision stands for the principle that, although Congress may repeal or amend the substantive provisions of legislation, it lacks the constitutional power to compel a particular judicial conclusion in a pending case, or a particular class of cases.

The language of the salvage logging rider is a prime example of the type of congressional intrusion that Article III and the *Klein* decision sought to prohibit. Without any reference to the substantive requirements or components of the relevant environmental laws (such as the National Environmental Policy Act, National Forest Management Act, and Endangered Species Act), the rider simply asserts that salvage logging satisfies these laws' provisions. The rider is nothing less than an explicit Congressional attempt to mandate a particular judicial conclusion to a specific class of cases. By deeming compliance in the legislation, Congress sought to preempt the Court's right to determine compliance in the courtroom.

Congress clearly has the constitutional authority to amend or repeal provisions of environmental laws, or any laws for that matter. Congress also has the constitutional authority to limit the application of laws, or to create legislative exemptions for a certain class of activities. Thus, Congress could certainly pass legislative amendments declaring that the National Environmental Policy Act, National Forest Management Act, and Endangered Species Act will not apply to salvage timber sales. The creation of such exemptions would be a constitutionally legitimate means for Congress to insulate salvage logging from environmental and wildlife protection restrictions.

To formally amend, repeal or limit these environmental laws, however, would likely be an explosive move politically. Such changes could not simply be tacked on as a rider to omnibus funding legislation in the manner that the salvage logging rider was inserted into the 1995 Recissions Act. They would have to work through the Senate and House committee process, and would therefore be subject to extensive hearings and public scrutiny. The environmental community would use such hearings and public scrutiny to full political advantage, targeting Congress members who proposed and supported these changes.

Given the potential political costs of formally weakening or repealing environmental legislation, it is understandable why members of Congress who supported increased logging were more inclined to take a legislative short cut. Instead of amending environmental laws, it was much easier to simply deem that actions which may violate the substantive requirements of laws were, nonetheless, still in compliance.

The public and the judiciary had been confronted with a similar situation in 1990, when Congress enacted the Northwest Timber Compromise. In this legislation, Congress declared that timber management plans in thirteen national forests in Oregon and Washington provided "adequate consideration" for habitat protection requirements under the Endangered Species Act.[3] The constitutionality of this "adequate consideration" language was reviewed by the *Ninth Circuit Court of Appeals in Robertson v. Seattle Audobon Society*.[4]

In *Robertson*, the Ninth Circuit found that the language of the Northwest Timber Compromise violated the separation of powers doctrine. The Court held that the "critical distinction, for purposes of deciding the limits of Congress's authority to affect pending litigation through statute, is between the actual repeal or amendment of the law underlying the litigation, which is permissible, and the actual direction of a particular decision in a case, without repealing or amending the law underlying the litigation, which is not permissible."[5] Finding that the "adequate consideration" language did not in fact amend or repeal any provisions in the environmental laws, the Ninth Circuit concluded that Congress had exceeded its law-making authority and intruded into the judicial arena.

The Ninth Circuit's *Robertson* decision strikes to the very heart of the constitutional debate over the salvage logging rider. It raises the question of whether Congress possesses the constitutional authority to simply deem that certain projects or activities comply with existing environmental legislation.

A Bad Decision Revisited

The Supreme Court's decision in *Klein* and the Ninth Circuit's decision in *Robertson* both point to the constitutional conclusion that the salvage logging rider violates the separation of powers doctrine. This conclusion, however, is weakened and possibly contradicted by a 1992 Supreme Court decision, which involved an appeal of the Ninth Circuit's constitutional interpretation of the "adequate consideration" language in the Northwest Timber Compromise.

In this 1992 decision, also entitled *Robertson v. Seattle Audobon Society*, the Supreme Court reviewed the Ninth Circuit's determination that the "adequate consideration" language violated the separation of powers doctrine. Writing for the Court, Justice Clarence Thomas set forth a rather novel interpretation of the Northwest Timber Compromise. According to Justice Thomas, the "adequate consideration" language did in fact amend several U.S. environmental laws, by permitting compliance either through these laws' existing provisions, or through compliance with the Northwest Timber Compromise.[6] Because Justice Thomas found that the environmental laws had in fact been amended, he concluded that "there is no reason to address the Ninth Circuit's interpretation of *Klein*."[7]

Although Justice Thomas's opinion is certainly a creative interpretation, it rests on some very thin and dubious jurisprudential grounds. Even if one accepts Justice Thomas' view that the Northwest Timber Comprise provided two ways of satisfying environmental laws, one of these ways still appears to run afoul under separation of powers. More specifically, the adequate consideration language still directs a particular finding—that the timber management plans are in compliance with all environmental laws. There is no new, or alternative standard articulated, which a court could then interpret or apply in conjunction with other previous environmental law provisions. Rather, there is simply the expression of a bald legal conclusion, that compliance exists.

More importantly, however, Justice Thomas's view should not be accepted because in fact Congress made no substantive changes to the law. The Congressional intent of the "adequate consideration" language was not to create a new standard for achieving compliance, but rather to prevent application of environmental laws to certain controversial timber management plans. As constitutional scholar Amy Ronner observed, "the Supreme Court focused not on what Congress actually did, but on what Congress *might* have done, because Congress *might* have modified the old laws instead of directing that compliance with new provisions satisfied the old laws . . . from *Robertson* emerges the notion that how Congress does what it did is virtually meaningless."[8] In short, the Court ignored the legislative language that was actually in dispute, and instead focused on the constitutionality of hypothetical legislation that Congress could have adopted. Because Congress could have amended the environmental laws at issue, the Court found that that was in fact what Congress had done.

Defenders of the Supreme Court's decision in *Robertson*, and of the "deemed to satisfy" language in the salvage logging rider, have also invoked the separation of powers doctrine to support their position. According to Mark Rutzick, a timber industry attorney, invalidation of the provision would represent judicial intrusion into the legislative arena, in that the courts would be improperly setting aside a democratically enacted law.[9] Rutzick maintains that, when read with other provisions of the salvage logging rider, the "deemed to satisfy" language simply clarifies that the law should be enforced notwithstanding any conflicting provisions in other laws. As such, Rutzick rightly points out that the main purpose of the "deemed to satsify" language was to excuse the rider from compliance with environmental laws such as the Endangered Species Act, National Forest Management Act and National Environmental Policy Act.

Yet, Rutzick's point is exactly the reason why Justice Thomas's opinion in *Robertson*, and the rider, are fundamentally flawed. Exempting a government action from compliance with a law is not the same as a bald declaration that a government action is in fact complying with, or satisfying, the substantive requirements of the law. The former is a valid political decision. The latter is legal conclusion, and in the case of the salvage logging rider, a critical misrepresentation.

The constitutional dimensions of the salvage logging controversy have not been lost on environmentalists. In the Fall of 1995, twenty two environmental groups from the United States, Canada, and Mexico filed a petition with the North American Commission on Environmental Cooperation (NACEC), established under NAFTA's environmental side agreement. The thrust of the NACEC petition was that provisions of the U.S. salvage logging rider blocked judicial review and therefore prevented effective enforcement of environmental laws. This, according to the petitioners, violated the terms of the environmental side agreement, which mandated that each NAFTA nation effectively enforce all environmental laws.

As Patti Goldman, an attorney with the Sierra Club Legal Defense Fund, stated in the NACEC petition: "The logging rider effectively suspends enforcement of environmental laws for logging of old-growth forests under Option 9 and salvage logging. For both logging programs, the rider provides that whatever environmental analysis is produced and whatever procedures are followed by federal agencies for such timber sales shall be deemed to satisfy the requirements of several specifically listed and all other applicable federal environmental and natural resource laws."[10]

The point raised in the NACEC petition is closely related to the constitutional argument. In short, legislative language that simply deems actions to be in compliance with the law blocks judicial enforcement and review. Under NAFTA's environmental side agreement, such language violates international law. Under separation of powers analysis, such language violates the U.S. constitution.

In response to national and international criticism, the Clinton Administration has taken steps to prevent, or at least limit, implementation of the salvage logging rider's "deemed to satisfy" language. On August 9, 1995, the Departments of Agriculture, Interior, Commerce, and the Environmental Protection Agency issued a Memorandum of Understanding (MOA) on timber salvage logging. In the MOA, the federal agencies pledged to conduct their activities in a manner "consistent with the National Environmental Policy Act and the Endangered Species Act" and to "implement salvage sales with the same substantive environmental protection as provided by otherwise applicable environmental laws."[11]

Although the Clinton MOA may provide an interim political solution, and may lessen the impact of the salvage logging rider on forest protection, the document does nothing to resolve the constitutional issues raised. The President's decision to refuse or limit enforcement, though a welcome development, merely creates a standoff between the executive and legislative branches. The MOA does not clarify whether citizens can still seek judicial review to ensure that salvage logging complies with federal environmental laws. The constitutionality of the underlying provision remains unchallenged.

As the NACEC petition and Clinton MOA indicate, it is time to revisit the Supreme Court's 1992 *Robertson* decision. By striking down the salvage logging rider's "deemed to satisfy" language, the judiciary can send a clear constitutional message that Congress will not be allowed to take political short cuts around the separation of powers doctrine. It would highlight that while Congress may adopt or change laws, it is ultimately the role of the courts to determine compliance or violations.

Moreover, this issue has implications far beyond the salvage logging rider. Increasingly, Congress is using riders in appropriation bills as a means to suspend or prevent application of environmental laws to specific projects. Thus, even when the Salvage Logging Rider expires, the separation of powers issues raised by the rider will not be moot. It is therefore critical that we reexamine, and reject, Justice Thomas's reasoning in the *Robertson* decision. This would place both constitutional law and environmental law on a much better course.

Chapter 3

Blaming Wildlife

The Endangered Endangered Species Act

If the conservative majority in Congress have their way, the newest political scapegoat may in fact be a goat. Or if not a goat, then maybe a bear, or a condor, or an owl. In the current anti-government congressional atmosphere, a powerful coalition of ranchers, developers and manufacturers has taken aim at the federal Endangered Species Act (ESA). This coalition, calling itself the wise use movement, maintains that one of America's chief impediments to economic growth is our national obsession with wildlife protection. The wise use movement maintains that the wisest use for our natural resources is maximum extraction, and that environmental policies must be rolled back to enable such usage and extraction.

Wise-users have chosen the ESA and wildlife protection as their political focus. Spearheaded by Rep. Don Young (R-Alaska), the chairman of the House Public Lands and Resources Committee, legislative efforts are underway to reverse the gains environmental protection advocates have made over the past twenty years. Despite evidence that most Americans favor strengthening environmental laws.[1] Young is pressing forward with a package of reforms that is likely to facilitate and encourage ecological deterioration.

Young's ESA reforms would allow the government greater leeway in authorizing the killing of endangered species and destruction of their critical habitat. Young would also give economists a central role in determining whether a species is facing extinction, by requiring the government to conduct cost benefit analysis studies before a species can be listed as endangered. Finally, the proposed reforms would require the government to compensate private landowners for not destroying critical habitat for endangered species.

In response to these congressional proposals, more than 150 environmental, Native American, scientific, business and labor groups have joined together to form the Endangered Species Coalition. Becky Dinwoodie, Northwest regional coordinator for the coalition, predicts that the debate over reauthorization of the ESA promises to be the conservation fight of the century.

Scapegoating the Owl

The wise use vs. wildlife debate has been propelled to center stage largely by the recent logging controversies in the Pacific Northwest. In lawsuits filed by the Portland Audobon Society, the Seattle Audobon Society and other environmental groups in the early 1990s, it was determined that the federal government issued logging permits in violation of the ESA and other environmental laws. These permits authorized logging of old-growth forests that were designated as critical habitat for the endangered spotted owl.

As a result of the litigation, the timber contracts were invalidated and logging was suspended until the impacts on the spotted owl could be assessed. To help resolve the Northwest Forest conflict, in 1993 the Clinton Administration helped negotiate the Option 9 settlement. The settlement was entitled Option 9 because the Clinton Administration rejected the eight settlement proposals put forth by the National Forest Service, deeming them too restrictive of logging, and instead developed its own ninth proposal. Although Clinton's Option 9 compromise convinced many of the environmental groups to drop their suits, this did little to dampen the animosity between the timber industry and environmentalists.[2]

Citing the mill closures resulting from the canceled logging contracts, wise-users launched a campaign to discredit wildlife protection laws. This campaign was based on two basic assumptions: first, that wildlife protection hurts the American economy; and second, that environmentalists vastly exaggerate the ecological threats posed by industry. Not surprisingly, these arguments found a receptive audience in those communities that were hardest hit by timber industry layoffs. Yet a critical look at the assumptions underlying the anti-ESA campaign reveals that owls—and environmentalists—are not to blame.

Wise-use advocates assert that strong environmental laws hurt the economy. At the Massachusetts Institute of Technology, political science professor Stephen Meyer decided to test this claim by assessing the impact of environmental regulations on state economies.[3] The MIT study found that states with the strongest environmental laws also have the strongest economies. States that neglect to protect their environments face long-term decline for a simple reason: The environmental irresponsibility of one industry negatively affects many others.

The Pacific Northwest timber industry is a prime example. The destructive clear-cutting of Pacific Northwest forests may have brought bountiful profits to the timber industry, but it has damaged many others.[4] Commercial fishing has suffered as rivers have filled with silt from the erosion of exposed hillsides. Tourism has also been hurt, since visitors have little interest in visiting clearcuts and stumps. In addition, taxpayers have often been forced to pay for the environmental restoration necessary to repair the damage caused by destructive logging. These cumulative losses often outweigh the benefits to one particular industry, resulting in overall economic decline.

This is why Oregon's economy continued to grow, and overall state employment increased, during the early 1990s.[5] To be certain, communities dependent on national forest timber were hurt by increased forest protection. As Stephen Yaffee, a forest policy expert noted, "These communities were on the front line in swallowing the economic impacts" of decreased logging.[6] These impacts, however, were inevitable as the region's timber base moved toward total depletion, as raw log processing jobs moved overseas, and as the timber industry moved toward greater mechanization. The economic downturn of logging-dependent communities thus began long before the forest protection battles of the early 1990s, and was likely to continue regardless of whether logging was permitted in the spotted owl's critical habitat.

Moreover, the depletion of the Pacific Northwest's timber base was due not to environmental laws, but rather to the U.S. government's and timber industry's mismanagement and over-exploitation of our national forests.[7] After generations of unsustainable logging, more than ninety-five percent of our nation's native forests have been cut down. This is why American timber companies are moving to British Columbia, Siberia and Southeast Asia and laying off American workers. Although the economic plight of timber-dependent communities is real, and needs to be recognized and responded to, we must remain clear about the true causes of their predicament.

ESA opponents also claim that the law has prevented government agencies from completing vital projects—such as dam, roads, and military cleanups. In fact, that hasn't happened. Between 1979 and 1991, more than 120,000 government projects were reviewed to make sure they were in compliance with the ESA. Thirty-four were terminated. This is less than .01 percent of all proposed projects.[8] Although a few notable projects, such as logging in spotted owl habitat, were temporarily delayed or modified, these delays and modifications have proven extremely rare. As the ESA plainly states, such measures are only required when a projects is "likely to jeopardize the continued existence" of a listed species.

The second wise use argument—that environmentalists trump up the dangers of all development projects—is equally suspect. While ranchers, developers and manufacturers maintain that threats to wildlife and species preservation are exaggerated, scientific evidence clearly indicates otherwise. In a 1992 study, Harvard biologist Edward O. Wilson concluded that, worldwide, more than 50,000 species become extinct annually, and that ten percent of all species now on the planet will likely disappear in the next twenty-five years.[9] According to Wilson, this is 10,000 times the natural rate of extinction. Moreover, of the hundreds of species listed as endangered or threatened under the ESA since 1973, most remain poised on the brink of extinction.[10] In fact, more species have become extinct than have recovered since the law was enacted.

The Nature of Reform

Although the wise-use critiques of the ESA are suspect, the law has been criticized even by environmentalists who acknowledge that species extinction is a real problem and that environmental protection and economic development are compatible goals. Much of this criticism has focused on the expense and inefficiency of the law's existing species-by-species approach. Under the current ESA scheme, separate habitat designations and recovery plans must be developed for each species listed as endangered by the U.S. Fish and Wildlife Service (FWS) and the National Marine Fisheries Service (NMFS). While these habitat designations and recovery plans are the most reliable way to ensure the survival of endangered species, they are scientifically complex and therefore expensive.

The FWS (a subagency of the Interior Department) and NMFS (a subagency of the Commerce Department) are not provided with adequate funds to fulfill their mandate under the current species-by-species ESA scheme. As a result, the vast majority of listed species have neither critical habitat designations nor recovery plans. According to a 1992 study by the General Accounting Office, critical habitat and recovery plans have been designated for only sixteen percent of the listed species.[11]

Given the Republicans' control of Congress, it is highly unlikely that the ESA's budget will be increased sufficiently to complete the remaining habitat designations and recovery plans. Many wildlife protection advocates have therefore proposed shifting the ESA's focus from "species habitat" to "ecosystem" protection. This approach was set forth in the Studds-Dingell Bill to amend the ESA. The bill, introduced by the House Merchant Marine Fisheries Committee in 1994, was, not surprisingly, dropped by the new Republican House leadership. The Studds-Dingell Bill called for recovery plans to be developed for ecosystems that contain a number of endangered species, rather than for each listed species. Ideally, this would result is more efficient use of administrative and scientific resources, and therefore cheaper and better species protection.

Although attractive to many environmentalists, the "ecosystem" approach presents many practical problems. Most importantly, the scientific definition of ecosystem is less precise and established than that of species. In the absence of hard science, politics and economics may play a large part in defining exactly what is a unique or endangered ecosystem. With a conservative Congress in power, this political latitude could result in a narrow definition of ecosystem and, therefore, less—not more—habitat protection for endangered species.

While reforms such as the ecosystem approach entail certain political risks, they should nonetheless be given serious consideration. Realistically, FWS and NMFS have little chance of receiving the funding they need to fulfill their ESA obligations under the current regulatory scheme. Streamlining the habitat designation and recovery plan formulation process may be the only way to ensure that species listed under the ESA are in fact protected.

Another, more recent, ESA reform proposal, calling for the creation of a habitat conversation trust fund, was proposed by Representative George Miller in 1997.[12] Miller's proposal focuses on improving habitat conservation on privately owned property. Under this proposal, property owners wishing to develop property that serves as habitat for endangered species may only do so if they post a performance bond. This bond, which is deposited in the habitat conversation trust fund, would be used to pay for additional habitat conservation if it becomes necessary in the future. The proposal is an attempt to provide some certainty to landowners, without locking the government into project approvals that could eventually lead to species extinction.[13] Although the proposed trust fund is supported by most environmentalists, some oppose it on the grounds that the amount of the performance bonds could define the extent to which the government may later protect endangered species habitat. As such, opponents of the trust fund view Miller's proposal as an unwarranted abdication of responsibility and authority under the Endangered Species Act.

Although ecosystem-based protection and the habitat conservation trust fund are not without their shortcomings, the proponents of these ESA reforms have at least identified the real issues at stake—unsustainable development and resource exploitation. As the Republican Congress turns its attention to the ESA, these issues must be kept at the forefront of the debate. The struggle to reform the ESA must not degenerate into blaming endangered species. Rather, it must focus on improving government policies and institutions so that the ESA can more efficiently protect wildlife and the environment.

Chapter 4

Words to Choke On

Free Speech and Environmental Debate

Under the free speech laws of both the United States and England, there are two areas in which expression has been permissibly restricted—in cases of defamation (libel/slander) and in commercial situations where information may be false or misleading to customers. Although the contexts in which these restrictions apply are quite different, both of these free speech limitations are based upon a common principle: that there are certain instances when the expression of inaccurate information is so damaging to individuals or society that public communication of such information should be illegal.

In San Francisco and London, two lawsuits have sought to apply these free speech restrictions in the environmental field. Unlike most environmental law cases, these suits did not involve hazardous waste liability, pollution or any such tangible natural resource issues. Instead, they focused on the right of individuals and corporations to interpret specific environmental terms. What the suits reveal is that the environment of expression, the free speech framework in which environmental discussion takes place, is now a critical legal question. The resolution of this question may determine not only the terminology, but the substantive scope, of environmental debate.

Paul Stanton Kibel

Misinformation in the Marketplace

In a 1992 suit in federal district court in Northern California, several national detergent, chemical, plastics and food industry groups sued the State of California.[1] In the suit, the industry groups challenged the constitutionality of California's 1990 "Green Seal" environmental labeling law.[2] This law was passed in response to complaints by consumer-protection groups that companies were making false and misleading environmental claims to sell their products.

The law's proponents cited such labels as "dolphin friendly," "made from recycled paper" and "ozone safe" as examples of situations where companies appears to be making certain objective product claims. The consumer-protection groups then demonstrated the discrepancy that often existed between the actual products and these apparently objective claims.

The Green Seal law sought to halt the potential for deceptive advertising by limiting the use of certain environmental terms in the commercial context. Under the law, companies are prohibited from making product claims such as "dolphin friendly," "made from recycled paper," "recycled," and "recyclable" unless the State of California certifies that certain objective, scientific criteria are met.

The suit brought by the industry groups alleged that, by seeking to outlaw corporate environmental definitions and control the public use of language, the Green Seal law violated free speech rights protected under the First Amendment of the U.S. Constitution. In support of their claim, the plaintiffs maintained that they were not merely trying to sell products. Rather, they argued, they were seeking to discuss policy issues with the public, and their interpretations were therefore entitled to full political protection.

The district court rejected the plaintiffs' attempt to invoke the broad protection of First Amendment political speech on behalf of corporate advertising. In its decision, the court clarified that the Green Seal law only applied in the context of product-related claims, and did not prohibit the plaintiffs from using the specified terms in other situations. Because commercial speech has never been afforded the same level of First Amendment protection as political expression, the law was upheld as a reasonable free speech restriction.

The district court's ruling was affirmed on appeal by the Ninth Circuit U.S. Court of Appeals.[3] The Ninth Circuit concluded that "merchant's commercial representations about the environmental attributes of their wares are far more likely to mislead consumers than editorial commentary opposing the statute or encouraging recycling and the use of biodegradeable materials."[4] The Ninth Circuit's ruling once again stressed that there are valid reasons for restricting free speech rights in the commercial context.

Facts and Fair Comment

In London, the McDonalds food corporation brought suit against two English environmentalists, David Morris and Helen Steel. As members of London Greenpeace, Morris and Steel distributed leaflets to customers outside numerous McDonalds restaurants in the London area. These leaflets accused McDonalds, among other things, of "wrecking the planet," and of encouraging tropical deforestation by raising cattle on recently deforested land in Central and South America. The leaflets added that "What's wrong with McDonalds is also wrong with all the junk-food chains. All of them hide their ruthless exploitation of resources, animals and people behind a facade of colorful gimmicks and family fun. . . . They are one of the worst examples of industries motivated only by profit and geared to continual expansion. This materialist mentality is affecting all areas of our lives, with giant conglomerates dominating the marketplace."[5] In response to these allegations, McDonalds sued Morris and Steel for libel, or as the British press took to calling the case, McLibel.

Because England does not have a constitution (at least not a single written Constitutional document), there is no exact equivalent to the First Amendment free speech protection provided under U.S. law. Instead, the English common law has developed a somewhat comparable doctrine called "fair comment" which provides citizens with considerable freedom to express subjective opinions. The fair comment doctrine, however, is generally considered narrower than U.S. free speech guarantees.[6] English judges are more willing than their American counterparts to scrutinize opinions or commentary to see whether they contain implicit or inferred factual claims. If such factual claims can be implied or inferred, and the claims cannot be substantiated, then the opinions or commentary may not be protected by the fair comment doctrine.

The McLibel litigation, which took over three years, from June 1994 to June 1997, was one of the longest libel trails in British legal history. According to the British legal press, McDonalds, which posted $24 billion in international sales last year, apparently believed that the threat of an expensive suit would silence Morris and Steel. Instead, it had the exact opposite effect.[7] Unable to afford a lawyer, Morris and Steel vowed to run their own defense. Moreover, the lopsidedness of the suit created a huge surge of international sympathy and support for the defendant-environmentalists.

To offer encouragement and financial assistance to Morris and Steel, McLibel support groups formed in Spain, Italy, New Zealand and the United States. Additionally, the former Assistant Attorney General of Texas, Steve Gardner, agreed to appear as a defense witness. Gardner testified that McDonalds consistently failed to comply with Texas consumer labeling laws and made health and environmentally-related product claims that were untrue.

In the suit, McDonalds sought to convince the court that what was at issue was not a matter of variant opinions, but instead a matter of truth and falsehood. McDonalds put on witnesses to defend and substantiate what it believes are environmentally responsible corporate practices. The company also called a nutritionist to testify about the health benefits of a diet based on hamburgers, fries and soft drinks. McDonalds asked the court to find that these practices were not merely arguably responsible or arguably nutritious, but responsible and nutritious in fact. This finding would provide English environmentalists with a strong incentive to keep their opinions to themselves.

On June 19, 1997, the McLibel case was decided. Although the court found that the plaintiffs' allegations of paying low wages, treating animals cruelly and targeting children in advertising campaigns constituted "fair comment," all other environmental allegations were found to lack a factual basis.[8] As a result of the ruling, David Morris and Helen Steel have indicated that they plan to take their case to the European Court of Human Rights, where they will likely argue that English libel laws violate civil liberties recognized under international law.

McDonalds' victory could have the practical effect of establishing corporate definitions of environmentally responsible behavior as new legal standards. Under the recent ruling, any attempt to characterize such behavior as irresponsible could expose the speaker or writer to damages for slander or libel. This new standard will likely have a chilling effect on environmental debate in the United Kingdom As Douglas Vick and Linda Macpherson, two English legal scholars, observed in a recent law review article, "The irony is that English defamation law deters critical reporting of precisely those whose activities most directly affect the public interest."[9]

Controlling the Debate

Taken together, the *Green Seal* and *McLibel* cases reveal that the relation between environmental goals and free speech is complex. In *Green Seal*, we find environmentalists arguing for commercial speech restrictions and corporate interests trying to invoke the political ghosts of Thomas Paine and Patrick Henry. In *McLibel*, we find corporate interests proposing that corporate definitions be adopted as legal standards while environmentalists defend their right to dissent and disagree.

Although the posture of the *Green Seal* and *McLibel* suits are different, in one important respect the cases are similar. At the center of both disputes there is a common recognition—namely, that public debate influences public opinion, and that public opinion influences corporate profits. This is why both environmentalists and corporations are taking the issue to the legislature and the courts. They both recognize the importance of controlling the terms of public debate about the environment.

The key question, however, is whether this control is likely to lead to more accurate information and a more vigorous debate, or whether it is likely to lead to misinformation and the suppression of debate. The public has a clear interest in ensuring that the former scenario prevails.

Chapter 5

Ignorance Abroad

International Projects Under National Law

For many years, the United States government has funded projects in foreign countries. This funding has focused primarily on countries in the developing world. Most of this funding has been targeted at economic concerns, such as road construction, natural resource development and industrial expansion. These projects, however, have also often had significant environmental and social impacts on their host countries, such as deforestation and pollution.[1]

Under the National Environmental Policy Act (NEPA), federal agencies are required to complete an environmental assessment before participating in major actions or projects that significantly affect environmental quality. While it is certain that U.S. funding of development projects abroad significantly affects the environment, it is uncertain whether such funding triggers the environmental assessment requirements under NEPA. The courts have not yet clearly resolved the issue.

There is one line of federal cases that have found NEPA applicable to foreign projects.[2] For instance, in the 1993 case of *Environmental Defense Fund v. Massey*, the Court of Appeals for the District of Columbia found that the National Science Foundation (a federal agency) must comply with NEPA's environmental assessment requirements before incinerating wastes in Antarctica. These decisions have allowed environmental citizen groups to challenge the government's failure to prepare an environmental impact statement in regard to international projects.

Another line of federal cases, however, has reached a contrary conclusion.[3] For instance, in the 1990 case of *Greenpeace v. Stone*, the District Court of Hawaii held that the U.S. Army was not required to comply with NEPA's environmental assessment requirement before transporting hazardous chemicals from bases in Germany to bases in the Pacific Ocean. These rulings have held that while NEPA requires the U.S. government to cooperate with foreign countries on joint international projects with environmental impacts, it does not mandate the preparation of an environmental impact statement.

The judicial debate over NEPA's application to projects on foreign soil has created confusion on both the environmental and diplomatic fronts. Environmentally, it has undercut efforts to ensure that the U.S. government does not promote resource degradation and pollution abroad. Diplomatically, it has left the executive branch uncertain what it must do to comply with U.S. environmental laws, and thus often unsure what foreign commitments it can make.

Relocating the Issue

The debate surrounding NEPA's application to foreign projects has so far focused on the location of the projects. The courts have generally assumed that because that because the projects are located outside the United States, NEPA will only apply if its environmental impact provisions were intended to have extraterritorial effect (apply outside U.S. territory). While this assumption appears logical at first glance, a closer look suggests that it is based on a critical misunderstanding of NEPA.

In focusing on the location of the projects and the question of extraterritoriality, the courts have overlooked NEPA's primarily domestic focus. At its core, NEPA is not so much about final consequences as about domestic procedure and accountability. The law's primary goal is to inject environmental considerations into government decision-making. These decisions are not foreign, international or extraterritorial. They are made in the United States by agency officials and involve the disbursement of U.S. federal funds.

The heart of NEPA is its environmental assessment process. Under this process, federal agencies are required to consider the likely environmental impact of proposed projects and to identify strategies to mitigate adverse impacts.[4] Because agencies may often have a vested interest in downplaying the environmental impacts of a proposed project, NEPA requires an open and transparent environmental assessment process. The federal government must make all pertinent project documents available to the public, and citizens and groups may submit comments and participate in agency hearings. Moreover, NEPA prohibits the "irretrievable commitment of resources" to a project before the completion of this environmental assessment. This prohibition recognizes that agencies are unlikely to alter or abandon a project already under way.

While NEPA's ultimate goal may be to promote environmentally responsible projects, its provisions focus almost exclusively on the promotion of an environmentally responsible decision-making process. It seeks to enhance the availability of environmental information to agencies and the public. The restrictions and procedures that NEPA establishes do not apply to projects (wherever they may be); they apply to U.S. government officials.

Lack of Recourse

When the U.S. government, through its funding decisions or projects, inflicts (or threatens to inflict) environmental damage on

U.S. soil, affected U.S. citizens have some avenues of political resource. First, they can pursue the democratic remedy of voting out the President and Congress that approved the project in question. Second, they can participate in NEPA's environmental assessment process and voice their objections and concerns.

These responses are not similarly available for damage inflicted on foreign soil. Foreign citizens cannot vote out the U.S. politicians who sponsored the project. Moreover, if NEPA is deemed inapplicable to international projects, affected persons will be unable to voice concerns and objections even at the project development phase.

While it may be argued that foreign citizens can seek redress from their own national governments, this response betrays considerable naivete. Developing countries, who are the recipients of most U.S. foreign funding, are frequently not open democracies. While, in theory, foreign citizens could appeal to their national governments, in reality, that political avenue is often closed.

The denial of NEPA's application to foreign projects creates a dangerous vacuum of political accountability and information. It results in the very situation that NEPA sought to avoid—decision-making by government officials unaware of or unconcerned about environmental consequences. It sets the stage for the export of ignorance, for the infliction of environmental damage on foreign citizens.

Consider the following foreign funding decision, an actual project being undertaken by the United States. To facilitate the economic development of the former Soviet Union (and to provide raw lumber for mills in the U.S. Pacific Northwest), the United States has developed a multi-million dollar project to help promote sustainable forestry in the Russian Far East.[5] Under this project, certain forests have been targeted for intensive logging. The project has also earmarked some funding for forest conservation.

In the Russian Far East, there is widespread concern that the proposed logging will result in U.S.-style clearcuts, with corresponding degradation of forests and wildlife habitat. Are there formal avenues for concerned Russians to voice these objections to the U.S. government? Are there formal procedures to ensure that U.S. decision-makers have accurate and comprehensive information about the environmental impact of the proposed project? Without NEPA's procedural safeguards, the answer to both of these questions is probably no. The goal of sustainable forestry may exist on paper, but there is little means of guaranteeing it will be achieved in practice.

Therefore, aside from the issue of extraterritoriality, denying NEPA's application to international projects also creates a troubling double standard for foreign citizens. As one legal commentator noted, "the United States is effectively stating that American citizens are more worthy of protection from environmental dangers than citizens of other nations. The environment cannot withstand this kind of governmental hypocrisy."[6]

Information and Responsibility

Extending NEPA's reach to international projects will prevent U.S. agencies from acting, to paraphrase Mark Twain, as ignorants abroad. So long as U.S. government officials remain unaware of the consequences of funding decisions, they will be unable to act responsibly. Information is the essential prerequisite to intelligent policy formulation, and NEPA can effectively guarantee that this information will be revealed and discussed. Those who believe U.S. projects will cause environmental harm can voice their concerns. Such information can only improve the funding and policy decisions of the U.S. government.

The United States, in all likelihood, will continue to play an active role in promoting economic development abroad. International projects must not inflict environmental damage beyond the territory of the United States. Such damage would not only injure the environment and long-term economic interests of the host country; it would also injure the credibility of the United States. The United States projects that harm the environment and citizens of other nations send a message to the global community that we are irresponsible.

The environmental assessment provisions of NEPA provide the best means for avoiding, or at least reducing the frequency of, such unfortunate funding decisions. Moreover, extending NEPA to international projects is consistent with the law's principal objective—namely to provide U.S. government officials with the information they need to make environmentally responsible decisions.

Part II.
FOREIGN SOIL

The American perspective on international environmental problems is rooted in the American experience. Our national politics and economy, and our geography and natural resources, provide the context in which we frame global environmental issues.

While the tendency to draw international policy conclusions from our own national experience may be understandable, it nonetheless results in a limited and often distorted analytical framework. The different legal traditions, political history and natural resource conflicts found in other nations dictate that the United States' model of environmental policy cannot be simply grafted on to foreign situations. Rather, environmental laws and institutions must be developed that draw on the existing traditions, and are responsive to the peculiar problems, within each nation.

This section examines legal and environmental issues outside the United States' borders. It analyzes how international environmental law is affecting the constitutional relationship between the federal and provincial levels of government in Canada. It considers how India's legal system is coping with the widespread death and injury resulting from the Bhopal tragedy. It assesses the difficult environmental questions confronting Russia and Vietnam as they seek to introduce market policies into their economies, and to revamp their legal systems.

In the environmental field, the international context is about more than abstract policy, and more than a global projection of the U.S. experience. It is about understanding the distinct resource issues, economic conditions, and legal traditions that underlie each country's national environmental policies.

Chapter 6

Axe to the Myth

Canadian Logging and International Law

Western Canada continues to inspire images of pristine, undisturbed wilderness. Perhaps the most dominant of these images is of the forests of British Columbia. These forests conjure up magnificent natural scenery. When one envisions British Columbia, one envisions temperate rainforests of western hemlock, cedar and Sitka spruce, its lush habitat for kermode bear and grey wolf, its clear rivers teaming with salmon. These places have taken on an almost mythical status, an expression of nature powerful and untouched.

Although powerful and compelling, the forests of Western Canada are in the midst of a transformation from reality to myth. The real forest with its real trees, real human inhabitants, real rivers and real fish is being clearcut at an alarming rate.[1] Clearcutting, the primary logging method practiced in Western Canada, is the most ecologically destructive logging technique known. It is also the most cost-efficient. Often described as strip-mining for trees, clearcut logging calls for the removal of all trees, plantlife and animals in a given area. All that remains is a wasteland of sticks and tree stumps.

Although Western Canada still contains large areas of undisturbed forest, these areas are now increasingly under threat. Clearcuts currently fragment most commercially viable forests in Western Canada.[2] In the coastal rainforests of British Columbia, most of the remaining old-growth forests are slated to be logged within the next two or three decades.

Although the clearcut of Western Canada is undoubtedly an environmental tragedy, it is also something else. It is a violation of international law and a breach of Canada's international commitment to sustainably manage its forests. In signing the United Nations Convention on Biological Diversity, the United Nations Statement of Forest Principles, the Migratory Birds Convention, the Pacific Salmon Treaty, and the North American Agreement on Environmental Cooperation, Canada has taken on certain international obligations.

Although Canada would prefer to frame the issue as one of domestic resource management, this framework must be resisted. In undertaking international forest protection obligations, Canada has acknowledged the global consequences of regional deforestation, and along with global consequences come global responsibilities. A review of the forests in British Columbia makes plain that Canada has failed to honor its international obligation to practice ecologically sustainable forestry.

Forests Under International Law

Canada's international forest obligations cannot be traced to a single document. Like many international duties, Canada's obligation to practice sustainable forestry is derived from several legal sources. Although each of these international legal sources approaches the issue of forest management from a different angle, common underlying principles can be readily identified. Among other things, these principles require that forests are managed in a manner consistent with their natural capacity for regeneration; that forests are managed to maintain and protect ecosystems; that forests are managed to protect critical habitat for birds and wildlife; and that forests are managed to preserve critical river and stream habitat for fish.

The Canadian federal government and the British Columbian provincial governments have argued that their current forest policies comply with these international principles.[3] According to these official sources, clearcutting is an "accepted practice," and is "entirely appropriate from an ecological standpoint for most forest types in Canada."[4] Moreover, these sources assert that Canadian forest policy sets a global standard for "wise stewardship" of natural resources. These contentions, however, are based on insupportable interpretations of such terms as "sustainable capacity for regeneration" and "ecological balance."[5] These interpretations have been categorically rejected by experts working in the forest management field.[6]

In sharp contrast to the interpretations promoted by Canadian government officials and the timber industry, the international scientific community has established standards for sustainable forestry practice. These standards emphasize the difference between sustaining the wood and timber supply, and sustaining the forest. As John Gordon, former Dean of the Yale School of Forestry, has noted, "The major change in forestry thinking has been the abandonment of the concept of a stable flow of wood from the land as a universally dominant management objective."[7] In place of the stable wood supply model, with its emphasis on the intensive logging of commercially viable trees, sustainable forestry calls for the protection and management of forest ecosystems.

Clearcut logging and the industrial forestry model currently advocated by the British Columbian government ignore these standards. Contrary to international standards, clearcut logging does not protect or maintain forest ecosystems. Instead, it results in wide-scale soil erosion, severe wildlife habitat loss, degradation of fisheries, and poor tree regrowth. In a 1994 report, even the British Columbia Ministry of Environment acknowledged the negative impacts of clearcutting. The 1994 report observed, "In lands managed for timber production, clear-cut logging, reforestation, and short rotations convert large tracts of mature or old-growth forests to managed forests, which do not support the same type of ecosystem as naturally disturbed forests. In effect, the natural forest ecosystem in such areas is permanently lost."[8]

The sustainable forestry standards articulated in international agreements should therefore be read in the context of science, not politics. The relevant environmental terms are not mere public relations soundbytes, to be interpreted to suit and justify current logging practices. They are identifiable criteria to be either respected or disregarded.

There are at least five international agreements that relate to forest practices and protection in Western Canada: the United Nations Convention on Biological Diversity, the United Nations Statement of Forest Principles, the Migratory Birds Convention, the Pacific Salmon Treaty, and the North American Agreement on Environmental Cooperation. Canada has signed and adopted all five of these agreements. The obligations set forth in these agreements, summarized below, help provide the basis for international sustainable forestry standards.

The United Nations Convention of Biological Diversity, or Biodiversity Convention, was reached in 1992 at the Earth Summit in Rio de Janiero. The agreement, which focuses on the conservation and sustainable use of the world's biodiversity, establishes at least three obligations that directly affect forest policy. First, the Biodiversity Convention requires that nations "regulate or manage biological resources important for the conservation of biological diversity whether within or outside protected areas with a view to ensuring their conservation and sustainable use."[9] Second, it demands that countries "promote the protection of ecosystems, natural habitats and maintenance of viable populations of species in natural surroundings."[10] Finally, the Biodiversity Convention calls upon nations to "adopt measures relating to the use of biological resources to avoid or minimize adverse impacts on biological diversity."[11] In the forestry context, this provision requires logging methods that avoid or minimize the destruction of critical habitat for endangered species.

The United Nations Statement of Forest Principles, also signed at the 1992 Earth Summit, is the first international agreement to focus exclusively on forest management practices. As one of its guiding principles, it recognizes the vital role that forests play in "maintaining ecological balance," and calls upon nations to protect "fragile ecosystems."[12] To assure that the world's forests are "sustainably managed,"[13] the agreement requires that countries strengthen "institutions and programs for the management, conservation and sustainable development of forests."[14]

The Migratory Birds Convention, an agreement among Canada, the United States and Mexico, was signed in 1916. Under the agreement, Canada is required to control development or resource use to prevent damage to the nests, eggs, and critical habitat of migratory bird species.[15] Several migratory birds covered by the agreement nest in forests being logged in British Columbia.[16] Under the Migratory Birds Convention, the Canadian federal government must ensure that provincial logging practices do not violate Canada's international obligation to protect the nesting habitat of these birds.

The Pacific Salmon Treaty, which entered into force in 1985, has two principle objectives—equity and conservation. The equity objective requires a fair and reasonable allocation of the salmon catch, particularly between Canadian and U.S fisherman. The fairness and reasonability of this allocation is based on where the salmon spawn. For example, there are salmon that are spawned in Canadian streams but then migrate into U.S. coastal waters, and there are salmon that are spawned in Alaskan streams and then migrate into Canadian coastal waters. The equity provisions are designed to ensure that each nation's catch corresponds with the number of salmon spawned in its streams.

The conservation objective of the Pacific Salmon Treaty focuses on preventing overfishing and habitat degradation, to protect the total salmon stocks in the Pacific.[17] One of the primary causes of habitat degradation is clearcut logging, which often results in severe erosion, river siltation and clogging of natural streams. The Pacific Salmon Treaty's habitat conservation provisions therefore creates a corresponding duty to avoid logging practices that destroy fish habitat. This duty was recognized by the Provincial Government of British Columbia in its 1995 report *B.C. Salmon Habitat*.[18] In this report, the B.C. government detailed the numerous forestry and watershed restoration projects the province has undertaken to fulfill its duty to conserve and enhance remaining salmon stocks under the Pacific Salmon Treaty.

The North American Agreement on Environmental Cooperation, or NAAEC, was signed in 1993. The agreement was negotiated because of concerns over the environmental impact of the North American Free Trade Agreement (NAFTA), and is therefore often referred to as the NAFTA environmental side agreement. NAAEC requires Canada, the United States and Mexico to effectively enforce their environmental laws and regulations and declares that it is inappropriate to encourage investment by relaxing environmental standards.[19] In Canada, the federal government, and the British Columbia and Alberta provincial governments, have adopted legislation and policies that require sustainable forest practices.[20] The failure by Canada to fully enforce laws mandating sustainable forestry constitutes a violation of the NAAEC.

The Fiction of Compliance

The international treaties that Canada has signed tell a pleasant story. They suggest that sustainable forestry is being practiced, biodiversity is being preserved and environmental laws are being effectively enforced. The real story of forest management in Western Canada, however, is quite a different tale.

In British Columbia, forest management continues to operate under industrial logging methods that have been obsolete and scientifically rejected for decades. Moreover, provincial governments have invested huge sums of public money in the timber industries they are charged with regulating. In this climate of outdated science and political collusion, Canada's international forest protection obligations have remained unfulfilled.

In British Columbia, the annual amount of forest logged on public land is more than twice the amount of forest logged in the national forests of the entire United States.[21] It is estimated that one thousand square miles of old-growth rainforest are felled in the province every year. Most of this cutting takes place away from main roads, and away from the eyes of travelers. Just beyond the scenic roadside corridors, however, lies a patchwork of vast clearcuts that extends for hundreds of miles along British Columbia's west coast.[22]

The ecological impacts of this cut rate have been predictably devastating. On Vancouver Island, for instance, less than seventeen percent of old-growth forests on flat or near-flat terrain remain.[23] Logging activities have therefore moved to steeper slopes, where the environmental damage caused is even more severe. This logging, and the road building associated with it, have caused serious erosion and landslides, with debris and sediment being washed into streams. In fact, only six of the Island's eighty-nine largest watersheds remain unlogged.[24] This has caused water degradation and severe damage to salmon runs.

A 1994 report by Environment Canada (the federal environmental agency) documented and criticized the environmental impact of B.C. logging practices.[25] The report warned that caribou and other large mammals were losing critical habitat, and that insects and small animals essential to the health of B.C. forests were being eradicated. Moreover, numerous indigenous bird species, including the white-headed woodpecker and great blue heron, are threatened by B.C. deforestation. The report also noted the poor health of second-growth trees, which appeared to be more prone to insect infestation and root disease than the old-growth forest they replaced.

The conclusions of the Environment Canada report have been corroborated by other sources. In 1992, the British Columbia Ministry of Forestry conducted an audit of fifty-four fish-bearing streams located near logging sites on Vancouver Island. The audit revealed that the thirty-four of the streams, nearly two thirds of those surveyed, had suffered ìmoderate to majorî damage.[26] In 1995, Raincoast Conservation Society (RCS), a B.C.-based environmental organization, released a study on the impact of logging on endangered grizzly bear populations. According to RCS, logging in coastal valleys has resulted in destruction and fragmention of grizzly bear habitat. RCS also report that erosion from clearcut logging has led to an acute decline in wild coho and sockege salmon populations, upon which the grizzly bear relies.[27] A December 1994 report by the Natural Resources Defense Council (NRDC) echoed these findings. The NRDC report detailed B.C.'s failure to incorporate basic environmental considerations into forest management policies. According to the NRDC study, B.C.'s logging "exceeds ecological sustainability" and "steep slope cutting practices have allowed such substantial soil erosion that regeneration may be impossible in some areas."[28]

The destruction of B.C.'s forests is due to more than just economics and outdated science. It is due to collusion between industry and government. The B.C. provincial government has directly invested public money in private timber and paper interests. In 1993, it purchased $50 million of stock in MacMillan Bloedel, the largest logging company operating in B.C.[29] In came as no surprise, therefore, when just weeks after the purchase,

MacMillan Bloedel was granted a huge logging concession on Vancouver Island.

Additionally, in the early 1990s MacMillan Bloedel and other Western Canadian timber corporations formed the B.C. Forestry Alliance, a public relations group whose purpose is to promote the industry's image worldwide. This group helped arrange B.C. Premier Michael Hardcourt's visits to the U.S. and Germany, where he sought to dispel the international criticism of B.C. logging practices. As B.C. journalist Joyce Nelson, who was recently honored by the Canadian National Association of Journalists, reported, "Quite literally, we are paying a few private companies to cut down our Crown forests and pocket the profits, leaving us with nothing but clearcuts and higher taxes. As if that weren't enough, our governments are adding insult to injury by paying for pro-industry propaganda abroad."[30] Given the ties between industry and the B.C. government, it is not surprising that environmental protections have often been unenforced and ecological warnings overlooked.

In response to widespread criticism, the B.C. government has taken some steps to improve forest management. In 1992, B.C. created an independent agency, the Commission on Resources and Environment (CORE) to help initiate a consensus-based process to resolve issues regarding logging and forest use. In 1994, B.C. adopted a new Forest Practices Code which reduces the size of clearcuts and limits logging near fish-bearing streams. Although CORE and the Forest Practices Code represent steps in the right direction, Canadian environmentalists maintain that they are of little practical benefit in that they set forth few binding, enforceable standards.[31] Hence, even with CORE and the Forest Practices Code, the B.C. government continues to promote widescale clearcut logging at the expense of ecologically sustainable forest management. The environmental principles set forth in these initiatives have not yet been effectively translated into fundamental policy changes.

Ottawa in the Woods

To its credit, the Canadian federal government has criticized forest management practices in British Columbia. Canadian Prime Minister Jean Chretien has even gone so far as to make a personal pledge to protect British Columbia's most threatened old-growth coastal rainforests, in Clayoquot Sound. Yet, despite these criticisms and pledges, Ottawa has not sought to directly interfere with provincial forest policies. Its position on the issue was made clear in a 1993 report by the federal Canadian Forest Service: "Forest management is a matter of provincial jurisdiction. Each province and territory has its own set of legislation, policies and regulations to govern the management of its forests."[32] As a result of this position, the federal government has refrained from applying federal environmental laws, federal forest policy or international treaty provisions to provincial logging.

Although Ottawa's reluctance to interfere can be attributed in part to the economic and political influence of the timber industry, there are also important constitutional considerations underlying its current position. Under the Canadian Constitution, powers are divided between the federal and provincial government. Unlike the division of powers set forth under the U.S. Constitution, the Canadian Constitution designates powers as exclusively federal or exclusively provincial. Textually, Canadian federalism does not allow for concurrent areas of legislation or regulation.

Although these black and white constitutional distinctions sound plausible enough on paper, they have proven extremely difficult in practice, particularly in the environmental field. This is because many environmental issues are inherently multidimensional and often implicate a number of sub-issues, such as health, agriculture, industry, commerce, labor, national security and foreign relations. This range of implicated issues has led to provincial and federal conflicts over jurisdiction.

The issue of forest management in the context of international obligations is an example of this constitutional tension. Under Section 94 of the Canadian Constitution, provinces are given exclusive power over property rights. Canadian court decisions have interpreted property rights broadly, to include issues of land-use and natural resources management.[33] This broad definition would appear to include forest management.

Under Section 91 of the Canadian Constitution, however, the federal government is given exclusive authority over "trade and commerce" as well as power to ensure "peace, order and good government" (commonly referred to as POGG). The Supreme Court of Canada has interpreted the POGG power to include matters of "national concern." Moreover, Section 132 provides the federal government with the power to directly implement international treaties concerning "trade and commerce" and POGG interests. The fulfillment of international environmental obligations would appear to fall within these enumerated federal powers.[34]

When applied to the implementation of international forest management agreements, the Canadian Constitution therefore points to two different interpretations. One interpretation is that because international forest management agreements concern property and natural resources, they fall under provincial jurisdiction. The other interpretation is that because international forest management agreements concern trade and commerce and are of significant national concern, they fall under federal jurisdiction.

The federal Canadian government has so far adopted the position that, under the Canadian Constitution, its hands are tied. This position was demonstrated most clearly in the NAAEC negotiations, when a special annex was created to enable the provincial governments to sign on to the agreement independently of the federal Canadian government.[35] This annex provided Ottawa with a short-term means of avoiding direct provincial-federal conflict over adoption of the NAAEC.

The current federal position, however, has been challenged by Canadian constitutional law scholars and the Canadian Supreme Court. These challenges cast considerable doubt on the legal effect of the NAAEC's special annex. They also raise a more fundamental question: Should the federal Canadian government be excused from compliance with international agreements in areas in which it is constitutionally competent to regulate?

In a 1991 law review article, "Federalism and Comprehensive Environmental Reform: Seeing Beyond the Murky Medium," Canadian constitutional scholar Rodney Northey considered the application of the Canadian paramountcy doctrine to federal regulation in the environmental field.[36] Like the supremacy doctrine in the United States, the Canadian paramountcy doctrine holds that when provincial and federal laws conflict, and both laws are valid exercises of jurisdiction, the federal legislation will prevail. According to Northey, this doctrine provides the federal Canadian government with constitutional authority to take a more active role in natural resources management.

The Supreme Court of Canada reached similar conclusions in the 1988 case of *Regina v. Crown Zellerbach* and the 1992 case of *Friends of Oldman River Society v. Canada*. In *Crown Zellerbach*, the Court considered the constitutionality of the Ocean Dumping Control Act, which provincial governments alleged went beyond the jurisdiction of the federal government. In upholding the act, the Court noted that, over time, issues can migrate from provincial to federal jurisdiction. Issues that were originally allocated to the provinces may evolve into matters of national and international concern.

In *Oldman River*, the Supreme Court considered whether the federal government could mandate environmental mitigation for a dam being proposed and funded by the Albertan government. Alberta argued that "the federal government was incompetent to deal with the environmental effects of provincial works." The Court rejected Alberta's position, stating that "although local projects will generally fall within provincial responsibility, federal participation will be required if the project impinges on an area of federal jurisdiction."

These holdings bear directly on the current debate surrounding the implementation of international forest protection agreements. The cases suggest that the real obstacle to federal action may not be the Canadian Constitution so much as a lack of political will. As a 1993 study by the Canadian Institute of Resources Law (CIRL) concluded: "The potential scope of the *Crown Zellerbach* decision is very broad, especially given the increasingly international focus on environmental problems. If the courts were to extend the rationale of the decision to other areas of international environmental concern, there is the potential for an increased—or at least different—federal role in environmental management of forests."[37]

Legal developments in other Canadian natural resources fields support the conclusion of the CIRL study. In the cases of fisheries and agriculture, constitutional conflicts over environmental jurisdiction have been resolved through what is called "cooperative federalism." Under this approach, Ottawa uses political and fiscal pressures to bring provincial policies in line with federal objectives. In practice, cooperative federalism has enabled the federal Canadian government to make significant jurisdictional inroads into areas that were previously provincial. These developments suggest that Canada has been moving increasingly towards a *de facto* (in fact), if not a *de jure* (in law), policy of concurrent jurisdiction in the environmental field.

In light of the paramountcy doctrine's potential applications, the *Crown Zellerbach* and *Oldman River* decisions, and the opportunities presented by cooperative federalism, the federal Canadian government would do well to reconsider its current position. While the loss of international credibility and the degradation of the global environment should provide incentive enough to provoke this reassessment, there is now an additional reason for

Ottawa to intervene. As a result of campaigns by international forest protection groups, corporate consumers of Western Canada's timber and paper products are cancelling their orders. In Great Britain, for instance, Scott Ltd. recently cancelled a $5.4 million contract with MacMillan Bloedel. Several newspapers, such as *The New York Times*, and phone book publishers, such as Pacific Bell in California, are also considering changing sources. These international developments should serve as a wake-up call to Ottawa. Like it or not, the clearcut of Western Canada is no longer simply an issue of national concern. As Steven Bernstein and Ben Cashore, political scientists from University of Toronto, noted in a 1996 paper: "Threats of foreign boycotts and international scrutiny have become commonplace for B.C. politics in the 1990s. . . . Policy actors, institutions, and economic forces from beyond the state have so affected B.C. forest politics that it no longer could be properly considered a domestic affair."[38]

Though slow to respond, the federal Canadian government has started to acknowledge the larger national and international implications of the problem. In 1995, Environment Minister Sheila Copps announced her support for the drafting and adoption of a federal endangered species act. Such a law could potentially provide a means to curtail provincial forest mismanagement. It remains to be seen, however, if Ottawa is willing to meet provincial opposition head-on. Will the proposed law provide the federal Canadian government with the means to ensure compliance with international obligations, or will it once again defer most substantive forest policy decisions to the provinces? Legislation that is rhetorically strong but substantively weak may provide Ottawa with good public relations material. It is unlikely, however, to end Canada's continuing violation of international forest protection agreements.

As Ottawa considers what course to take, it might be useful to consider the following hypothetical question. What if the State of New York announced that it had no intention to abide by U.S.

obligations under NAFTA? Most likely, Canada would demand that the U.S. federal government intervene. Confronted with British Columbian violations of international environmental law, the international community must make this same demand. For the sake of the global environment, and for the sake of its own international credibility, Canada must find a way to put its house in order.

Chapter 7

Ecology after the USSR

Hard Times for Russian Environmental Law

The collapse of the U.S.S.R. was both a consequence of, and a catalyst to, the ecological awareness of its citizens. The introduction of *perestroika* in the late 1980s was accompanied by the emergence of a vocal and politically potent environmental movement.[1] Citizens began to discuss the environmental problems afflicting their country, and these discussions led eventually to criticism and protest. These criticisms and protests addressed a broad range of issues, including air and water pollution, deforestation, soil erosion, nuclear energy and public access to information. The environment served as an important focal point for citizens working to reform, or in some cases dismantle, the political structure of the U.S.S.R.

The disintegration of the U.S.S.R has resulted in a transfer of primary political power to the new independent republics. The Soviet disintegration has also resulted in an expansion of civil liberties, particularly those of free speech and free press. This decentralization and liberalization would appear to provide a good foundation for addressing the consequences and causes of environmental degradation. Such a response, however, has not yet been forthcoming.

Despite the emergence of new independent states, and despite the adoption of new laws, the environmental deterioration of the former Soviet Union has continued. This deterioration is particularly evident in Siberia and the far eastern regions of Russia.[2] When Russia achieved independence in 1991, it obtained control over the vast forests, rivers and natural resources in Siberia. Because of its valuable timber and petroleum reserves, the Soviet Union had devoted considerable energy to the extraction of these resources. This devotion, however, left Siberia with a host of environmental problems. With its forests and wildlife disappearing and its air and water quality standards declining, many looked to the new political regime to improve these conditions.[3]

Although there are now greater environmental protections on the Russian law books, these protections have so far done little to improve ecological conditions in Siberia. Despite the significant restructuring of political power and the increasingly open discussion of ecological issues, the Siberian environment continues to deteriorate.

The state of the region's environment is due in large part to the economic and legal legacy that Siberia and Russia inherited from the U.S.S.R. This legacy helps explain why the dramatic legal and political changes in Russia have not resulted in more effective nature conservation and environmental protection policies. Confronting and overcoming this legacy is one of the most critical tasks now facing environmental advocates in Siberia and Russia.

The State of the Environment

Siberia is endowed with natural resources of significant ecological and economic importance. In assessing the present condition of the Siberian environment, it is necessary to keep in mind the relation between the natural and the financial. It is only through this co-mingled assessment that an accurate portrait of the Siberian landscape can be presented.

Siberia contains one-fifth of the earth's forest cover, and one-half of the earth's coniferous forests. These forests cover an area roughly the size of the continental United States. They serve a number of critical ecological functions.[4] First, they remove huge quantities of carbon, perhaps as much as forty billion tons, from the atmosphere. The Siberian forests, therefore, play an important role in maintaining the world's climate balance. Second, they are habitat for several threatened and endangered species, such as the Siberian Tiger and the Far Eastern Leopard. The survival of these species depends, in large part on the conservation of their natural forest habitats. Third, they provide cover for watersheds and prevent soil erosion, which adversely impacts water quality. These watershed and anti-erosion functions are necessary to ensure adequate river flows as well as the health of fisheries and potability of drinking water.

In spite of their ecological significance, the Siberian forests have been targeted for intensive logging by domestic and foreign timber companies. These companies employ cutting techniques that violate accepted standards of ecologically sustainable forestry. In the Kirzhinsky region of Siberia, clear-cutting of forests has caused widespread soil erosion and siltation. As a result, several rivers in the region have suffered from severe siltation, and some tributaries have filled in and disappeared. This has damaged fisheries and reduced fresh water resources. The use of out-dated logging and transportation methods results in the vast majority of cut wood never making it to market. It is estimated that close to fifty percent of the wood cut by the Russian timber industry is left to rot.

In addition to the direct consequences of the logging, the Siberian forests are also severely impacted by pollution from oil, gas, and coal extraction. The condition of the forest is further compounded by the harsh climatic conditions in Siberia, which make for low metabolism and thus slow tree recovery. The result has been an accelerated decline in both the vitality and coverage of the Siberian forests, and an accompanying decline in global carbon absorption, endangered species habitat, soil conservation, and water quality.

Beyond its valuable forests, Siberia also contains large deposits of minerals and petroleum resources. Seventy-five percent of all Russian silver is obtained through the production of polymetallic ores in East Siberia. The region supplies a significant portion of the nation's gold and nickel. Siberia is also home to some of the largest oil, natural gas, and coal reserves in the world. In the former Soviet Union, extraction and processing of these mineral and energy resources was a top economic priority. Roads and railroads were built to obtain and transport the resources. Factories were constructed to process and store the resources.

Because extraction operations and factories were designed with minimal pollution controls, these industries have devastated large portions of the Siberian environment.[5] Oil invades lakes and rivers and seeps into underground aquifers. Gas flares and black clouds from burning waste pits cover the landscape, raining poisonous soot on the surrounding trees and plants. The forests are littered with abandoned machinery and rusting pipes.

The pollution and ecological damage caused by the Siberian mineral and petroleum industries affect more than lakes and plants. They affect the health of Siberia's human residents as well. In industrial Siberian cities such as Bratsk and Noyabr'sk, air and water pollution have resulted in severe health problems, including increased cancer rates and respiratory disorders among children.

Pollution is also poisoning Siberia's Lake Baikal, the deepest, and one of the largest, lakes in the world. To process the raw logs that are being cut in the surrounding forests, lumber and paper mills have been constructed on the shores of the lake. The mills contain minimal or no pollution controls, and thus large amounts of untreated waste water are discharged directly into Lake Baikal. This pollution threatens to destroy the fragile ecosystem of this unique and globally significant natural resource.

The State of the Economy

Although Siberia has long been a source of valuable timber, mineral and petroleum resources, these resources have generated little to improve its citizens' standards of living. The failure is due in large part to the political and geographical dynamics of the former Soviet Union. While the natural resources may have been located in the Siberian east, the political and economic power remained in the west, in Moscow. Under the highly centralized U.S.S.R. regime, the planning and funding for Siberian resource exploitation emanated from the political center. It is therefore not surprising that most of the wealth generated from these investments returned to the west, to the center of economic and political power.

The collapse of the Soviet Union has loosened Moscow's control over Siberia's natural resources. This loosening of Moscow-based control has created a considerable vacuum of political power in Siberia. In this vacuum, local and regional governments have asserted control over natural resources. Given the absence of clear laws governing the privatization of formerly state-owned land, these local and regional governments have been dispensing land and resource rights in a legal and regulatory void. This confusion over property rights, privatization and federal-regional jurisdiction has in turn led to the rise of what local citizens call the "timber mafia" and other corrupt natural resource syndicates. Many of these resource syndicates have close ties to local and regional governments. The majority of profits generated by these groups do not return to the local Siberian economy. Instead, it is widely presumed that the monies are transferred to protected foreign bank accounts. Siberians thus find themselves fighting a familiar battle. Even with the collapse of the U.S.S.R., the economic benefits of its resources continue to flow outside the region.

The liberalization of foreign trade in Russia has added a global dimension to this economic pattern. The planning and funding for Siberian resource exploitation now also comes from foreign corporations. Timber companies from South Korea and the United States, and mining companies from Canada, have negotiated deals with regional governments in Siberia. The priority of these foreign corporations is to extract and transport Siberian natural resources as inexpensively as possible. This priority means that wages and environmental standards remain low. It also means that most of the profits will be returning to foreign corporations and shareholders, not the local Siberian economy.

Siberia's economic condition has also been adversely impacted by Russia's program of demilitarization. Just as in the United States, in Russia the end of the cold war has meant a reduction in military budgets. Under the former Soviet Union, Vladivostok served as the home of the Pacific Naval Fleet. As part of the program of national demilitarization, the Vladivostok naval base was drastically downsized, and several other Siberian military bases were closed. Converting from military to civilian industries has added to Siberia's economic woes.

The dire economic situation of most Siberians is closely related to the continuing deterioration of the Siberian environment. The flow of profits outside the region and the economic displacement caused by the downsizing of the state-military complex have resulted in great hardships. These hardships, if not addressed, threaten to overwhelm efforts at ecological reform.[6] Efforts to improve ecological conditions in Siberia are therefore inevitably linked to improving the economic condition of the Siberian people.

A Difficult Legacy

Seventy-four years of Soviet rule have left Siberia, and Russia as a whole, with a problematic legal legacy. This legacy is problematic because it lacks many of the legal and political traditions that would help further the establishment and operation of a modern democracy. Because any ecological reform will take place in the larger context of democratic and political reform, the absence of these traditions is a considerable obstacle to improving environmental conditions in Siberia. In particular, there are three legal legacies of the Soviet era that hinder efforts at ecological reform—the disparity between law on the books and law in practice, the absence of an independent judiciary, and the absence of environmental considerations in Soviet economic development. These three legacies, discussed in greater detail below, need to be reckoned with before serious progress can be made on the environmental front.

First, in most modern democracies there exists a basic political assumption that after a law is enacted, the law will be implemented. It makes little difference whether the law is unfair, poorly drafted or serves particular interests. There remains a basic belief that the government will give effect to the provisions of the law. Because of experiences under the former Soviet regime, this underlying expectation is not part of the Russian legal tradition.

In the former Soviet Union, laws were routinely adopted and then ignored. Sometimes the lack of implementation was a product of political intent; other times it was a result of inadequate human or administrative resources. Regardless of the reasons for the lack of implementation, the end result was a situation where law in practice bore little relation to law on the books. Much of the U.S.S.R. was governed by what one commentator called "lawlessness law," wherein "who you knew was more important than what the rules said and where most rules were not really rules as all, but guidelines for action, subject to endless variation when applied in practice."[7] This low regard for the written rule of law was not surprisingly accompanied by a lack of respect for the legal profession.

This skepticism regarding the operative effect of the written law suggests that passing new laws will not be enough. Special efforts will have to be made to ensure that provisions enacted are indeed implemented, or at least capable of being implemented. If not, the tradition of "lawlessness law" will unfortunately be carried over into post-Soviet Russia.

Second, an independent judiciary is an essential political component of almost all modern democracies. This is particularly true of the United States and other countries whose legal systems derive from the English common law. Under the common law system, courts and judges retain considerable autonomy and law-making power. Even in countries which operate under the civil law system, however, such as France and Germany, courts play an important role in the application of law. The courts in the civil law system still provide a political forum for both the government and private parties to assert legal violations. Moreover, they possess the necessary enforcement powers and political standing to ensure compliance with judicial rulings.

In the former Soviet Union, the judiciary never secured the status, autonomy, or enforcement powers necessary to play a meaningful independent role.[8] Judges were appointed and removed at the absolute discretion of Communist Party leaders. Standing to sue was routinely denied to citizens challenging the actions of state officials. Judicial determinations that conflicted with other government priorities were simply not enforced. Given the minor role played by the courts in the former U.S.S.R., the post-Soviet judiciary in Russia begins as a relatively weak institution. Although the political regime has changed, the courts are still widely perceived as ineffectual.[9]

This perception of the judiciary adversely affects the implementation of Russian environmental laws. Although prohibitions and guarantees exist on paper, the courts must provide a legal forum when these prohibitions or guarantees are violated. If the courts cannot, or will not, provide this forum, the judiciary will retain its low status, and the disparity between the written and the actual law will continue.

The third unfortunate legacy is the absence of environmental considerations in Soviet economic planning. Beginning in the

1920s, the Soviet Union made industrial and agricultural modernization its top economic priority. The central government set specific regional goals, and local officials were responsible for ensuring that these targets were met. The targets commonly set forth manufacturing quotas or levels of agricultural yield. Failure to achieve these levels of production and output were often viewed as crimes against the state. As such, local officials had significant personal incentives to make certain that targets were met.

These economic targets often failed to account for ecological limitations. Forests were felled at rates that did not allow for regrowth. Factories were constructed with little or no pollution controls, rendering air and water poisonous. Intensive farming practices were employed which outstripped the soil's regenerative capacity. Many Soviet scientists and politicians were aware that the economic targets were not sustainable from an environmental standpoint. The political climate of the U.S.S.R., however, where questioning official policy was a punishable crime, made open discussion of the incompatibility between economic models and ecology a potentially dangerous undertaking.

Under the Soviet regime, citizens and regional officials were given little opportunity to incorporate environmental considerations into policy decisions. They were also prohibited from openly discussing the ecological consequences of these decisions. As Russia moves toward a more open economic and political system, and ideally more environmentally sustainable development, the lessons of complacency and silence will have to be unlearned.

On the Page and On the Ground

The collapse of the Soviet Union has created a political vacuum in Siberia. Moreover, because of the legal legacy of the U.S.S.R., Russians are accustomed to law in practice bearing little relation to law on the books. This chaos and lack of confidence in existing written laws underlie present legal efforts protect the Siberian environment. An understanding of the federal and regional environmental laws therefore requires that one look beyond the legislative texts and consider how these laws operate in practice.

At the federal level, Russia has adopted three laws since the collapse of the U.S.S.R. that bear heavily on ecological and conservation issues: the 1991 Land Code of the Russian Republic, the 1991 Federal Act on Protection of the Environment, and the 1993 Russian Forestry Act.

The 1991 Land Code represents an attempt to move away from the central government's monopoly of land ownership under the former Soviet Union. The code has been characterized by one commentator as the "rebirth of a dormant concept of private ownership."[10] Under the new code, land is characterized according to the purpose of the activity performed on the land. The categories include farmland, urban, industrial land, recreational and conservation land, forest land, and water resources land. These categories are relevant in that they determine the extent to which the federal government retains control over the land, as well as the degree to which privatization is permitted.

The Land Code provides for the establishment of local councils of People's Deputies. These local councils are in charge of granting land parcels. They are empowered to grant ownership, an inheritable life-long possession, a permanent or temporary use, or a lease on the land. The local land councils' disbursement powers are limited by the type, or category of land in question. If the land involved is to be used for agriculture or industry, the land councils have virtually absolute discretion. If, however, the land is to be used for recreation or conservation purposes, then the local councils may only allow limited business activity. Moreover, these lands may not be disbursed in a manner that adversely affects these primary recreation and conservation purposes.

The land disbursement scheme established under the 1991 Land Code has proven faulty in practice for two reasons. First, the composition and selection of the local land councils was not clearly set forth in the law. This has resulted in political uncertainty as to who may sit on the councils, and competing claims of legitimacy between different local councils. Second, land is categorized not so much by its nature as by its use. Local councils can therefore often avoid the conservation and recreation restrictions by recharacterizing the land's use as agricultural or industrial.

Competing claims of legitimacy and manipulations of the Land Code's environmental restrictions could, theoretically, be resolved through judicial clarification. In reality, however, this has not happened. The Russian courts presently lack either the interpretational or enforcement powers necessary to resolve these conflicts. Without a legal forum to settle these disputes, the Russian privatization process has been rife with corruption. Given its geographic remoteness from Moscow, this situation has been particularly acute in Siberia.

The 1991 Federal Act on Protection of the Environment was enacted to provide a comprehensive legal framework for environmental law. The law sets forth two primary legal regimes. First, it established the requirements for an *expertiza*, the impact report a government agency must prepare whenever a proposed action could adversely affect the environment. Second, it provided for the creation and management of national parks and reserves.

The *expertiza* requirements set forth in the Russian Federal Environment Act are similar in many respects to the Environmental Impact Statement (EIS) requirements set forth in the United States' National Environmental Policy Act (NEPA). Like the EIS under NEPA, the Russian *expertiza* must discuss the foreseeable adverse environmental effects of a given projects, as well as what steps can be taken to minimize these effects.

Unlike NEPA, however, the Russian *expertiza* does not include mechanisms for public participation in the preparation of the impact statement. Because the government often has an interest in concluding that the adverse environmental impacts of a project are minimal, this lack of openness undercuts the credibility of the *expertiza* process. Moreover, regional governments also face considerable anti-*expertiza* pressures from the general public, who bear the tax burden for carrying them out. Regional governments will thus often reach the convenient preliminary conclusion that no adverse effects are foreseeable, and that therefore no *expertiza* is required.

The national park and preserve regime established under the Federal Environment Act vests authority for the management of these areas solely with the Ecology Ministry of the Russian Federation. Within established park and preserve zones, development is permitted only so long as it is consistent with the ecological maintenance of the protected area. Although this regime appears rational on paper, it has not been provided with adequate funding and has therefore failed to designate the boundaries for many protected areas, or establish criteria for creating such protected areas in the future.

The Russian Forestry Act, adopted in December 1993, revised existing forestry laws in several respects. Unfortunately, these revisions appear to weaken rather than strengthen forest protection.[11] Under previous forestry laws, citizens and non-governmental organizations were given express authority to participate in forest management decision-making. This right to participation was secured through the agency inspectorate, which solicited and submitted to forest agency officials the comments of interested parties. This comment and criticism process has, in recent years, involved thousands of individuals and groups throughout the country. In Siberia, where ecological mismanagement of forests has been particularly severe, this process served as an important vehicle for environmentalists to voice their concerns. Although these concerns may have often been overcome by short-term economic interests and corruption, the existence of legally recognized forum for debate was nonetheless important.

In the new Russian Forestry Act, the provisions recognizing the inspectorate was removed. As a result, it is uncertain what role citizens and environmentalists will play in forest management decision-making.

An additional, and potentially positive, change in the Forestry Act concerns the clarification of federal and regional jurisdiction over the forests. Under the new act, the federal government retains exclusive authority over all national parks and reserves. Moreover, no commercial logging would be allowed in these federally protected areas. As such, these provisions parallel and expand on the park and preserve provisions set forth in the 1991 Land Code and the 1991 Federal Environment Act. The prohibition on commercial logging in effect defines the forestry management practices consistent with the areas' primary recreation and conservation purposes.

Unfortunately, the Forestry Act also suffers from the same defects as prior laws. Commercial logging may be prohibited in federal parks and preserves, but the federal government has not yet clearly established the boundaries of, and jurisdiction over, these lands. In the absence of such boundaries and effective federal jurisdiction, regional governments will likely continue to assert control over these areas, and the destruction of the forests will continue.

At the regional level, environmental protection efforts have fared little better. Under the 1991 Land Code, 1991 Federal Environment Act and 1993 Forestry Act, *subrepubliks* (regional governments) are given considerable authority to adopt laws that protect the environment. The Land Code grants *subrepubliks* the power to control privatization and land-use outside of federally recognized area. The Federal Environment Act provides that *subrepubliks* can adopt laws forbidding private operations that adversely impact the environment. The Forestry Act permits *subrepublik* regulation on all non-federal forests. Taken together, these laws help establish the framework for a truly federal system, wherein regional governments retain considerable power.

In the long run, the emergence of this federal framework holds great possibilities for Siberian environmental protection. In the short run, however, it has led to increased environmental degradation. Poverty and political chaos at the regional level have created a situation not conducive to ecological reform. Siberian *subrepubliks* are currently more concerned with the business of reducing employment and attracting foreign investment, and environmental protection is not at the top of their policy list.

The situation at the regional level once again highlights the need to integrate environmental reform with strategies for economic development. Although poorly planned economic development is often the cause of environmental damage, the complete absence of economic activity can also contribute to the problem. When people are focused on securing food, shelter and clothing for their families, the environment is not likely to receive serious attention. Therefore, the solution is not to stop all economic development, but to create economic development that reinforces and actually contributes to environmental protection. Local economies must be developed that strengthen and protect communities and ecosystems, not damage and exploit them.

Solutions Beneath the Surface

The ecological deterioration of Siberia is due in large part to the legal legacy left behind by the Soviet Union, and the difficulties associated with implementing environmental laws. Although the passage of new environmental legislation may have some cosmetic appeal, such laws are unlikely to improve Siberian ecological conditions unless these underlying legal issues are confronted. Two strategies that may help address these underlying problems are the expansion of citizen enforcement provisions and the clarification of the land privatization process.

First, the expansion of citizen enforcement provisions recognizes that adopting a law does not ensure that a law will be implemented. The implementation of a law can only be guaranteed by providing specific procedures to prosecute and punish violators, be they private corporations or agency officials. Without such provisions, citizens, corporations and the government have no clear incentive to obey the law. Without such provisions, written law and law in practice begin to diverge, and the rule of law begins to break down.

This situation becomes particularly acute in political systems, such as Russia, that lack a strong, independent judiciary. In such political systems, judges lack the power to create doctrines that would allow for the prosecution and punishment of violators. Also, the courts cannot be relied upon to supplement the textual provisions of a law so that implementation is possible.

Because a strong independent judiciary cannot be created overnight, Russians wishing to ensure government and private compliance with environmental laws should push for citizen enforcement provisions in legislation. These provisions should be highly specific. They should provide that any person or organization whose interests are potentially threatened may initiate a suit for compliance. They should also expressly grant courts the power to fine and imprison individuals who knowingly violate the law, as well as the power to enjoin activities that violate the law.

Strong and highly specific citizen enforcement provisions would serve many important functions. The necessary prosecution and punishment components to ensure proper implementation of existing environmental laws would be in place. This, in turn, would help reduce the divergence between written law and law in practice. By merely applying the express language of these citizens enforcement provisions, judges would improve the credibility of the courts.

Second, clarification of the land privatization process should be a policy priority. In considering the relative ineffectiveness of Russian environmental legislation, uncertainty regarding the ownership of and jurisdiction over real property has merged as a constant problem. The 1991 Land Code established a legal regime for categorizing land and a process for gradual privatization. The categories and privatization process set forth in the Land Code in turn formed the basis for subsequent distinctions in the Federal Environment Act and Forestry Act. Because of this legislative progression, the proper functioning of the privatization process is essentially a prerequisite to the implementation of these subsequent laws.

Unfortunately, as detailed above, this privatization process has so far been unsuccessful in clarifying the legal status of Russian land. The privatization process was carried out by local land councils whose legitimacy was often unclear. These councils were then required to make distinctions based on categories of land which the federal government has not yet clearly identified. The legal results of this process have been understandably ambiguous.

Before existing environmental protections can be implemented, or even made comprehensible, Russia will have to return to square one. The privatization process must be revamped. Until this is done, Russian environmental laws will remain incoherent, and thus incapable of implementation.

Experience and Expectations

The collapse of the U.S.S.R. has provided Russia with an opportunity to stop the ecological deterioration of Siberia. The realization of this goal, however, requires more than the adoption of aspirational environmental laws. It requires a critical reassessment of the reasons underlying Siberia's flawed ecological-legal framework.

The starting point for this reassessment should be the legal legacy inherited from the U.S.S.R. This legacy accounts for many of the shortcomings and defects of existing environmental legislation. Efforts must therefore be made to unlearn the lessons of the Soviet era. The judiciary must establish itself as a political forum where violations can be asserted, and violators punished. Procedures must be developed that allow citizens to help enforce environmental laws. More significantly, however, Russians and Siberians must come to expect that law in practice will reflect law on the books. Only then will ecological protections move beyond the page, and into the forests and rivers of Siberia.

Chapter 8

United By Poison

Relief for Bhopal's Victims

The Bhopal disaster, which took place just after midnight on December 3, 1984, is one of the worst industrial accidents in history.[1] Forty tons of highly toxic methyl isocyanate, which had been manufactured and stored in Union Carbide's chemical plant in Bhopal, India, escaped into the atmosphere. The accident killed over 3,500 people who lived in the dispersing chemical's pathway. Over 300,000 were injured—many seriously and some permanently.[2] These figures, as disturbing as they are, however, do not convey the full scope of the tragedy—the families destroyed, the communities impoverished, and the land poisoned.

In 1986, after unsuccessfully seeking to sue Union Carbide in federal district court in New York (where Union Carbide was incorporated), the Indian government filed suit against the multinational corporation in Bhopal District Court, India. The district court's ruling was eventually appealed to the Supreme Court of India. While the case was still pending before the Supreme Court, the Indian government and Union Carbide reached a settlement.

The settlement required Union Carbide to pay the Indian government $470 million (if the case had been decided in the U.S. under New York tort law, it is estimated that the Indian government could have recovered close to $4 billion). This settlement would be supervised by the Indian Supreme Court, and distributed by special compensation courts, to those who had been injured, and the surviving families of those who had been killed, in the Bhopal disaster.

The Bhopal settlement distribution scheme devised by the Indian Supreme Court was based primarily on compensating specific individuals for death or injury. This individual-based distribution scheme, although well-intended, has failed to respond to the severe medical and social consequences of the Bhopal disaster. To respond to these problems, the existing individual-based scheme needs to be integrated with community-based distributions. These community-based distributions would fund institutions, programs and services that serve the larger collective group of persons poisoned and injured by the Bhopal accident. This, in turn, would help provide more effective long-term relief for survivors.

A Blanket Spread Too Thin

Under the Bhopal settlement distribution scheme set forth by the Indian Supreme Court, 30,000 individuals were expected to receive compensation payments. These individual payments were intended to cover pain and suffering, lost wages, and incurred and future medical bills. The amount of compensation that each individual was to receive would be determined by the severity and permanence of the individual's injury.

Although originally 30,000 claimants were expected, the class of claimants has ballooned to over 300,000. Because the total settlement is fixed, this enlargement has drastically reduced the amount of money each individual will receive. This enlargement has also drastically increased the administrative burden placed on the compensation courts and government officials.

While the individual-based compensation scheme appeared as a fair and effective response when it was first established, it is now clear that the scheme needs revising.[3] In addition to the drastic reduction in the amount of individual payments (due to the enlargement of the class of claimants), there are other ways in which the scheme has proven flawed. These flaws have prevented the Indian government from achieving basic medical-, social- and justice-related goals. The most critical shortcoming of the current compensation regime is its failure to provide for the future of the Bhopal community and affected unborn generations.

The Bhopal disaster placed forty tons of highly toxic methyl isocynate into the Bhopal atmosphere. In addition to the death and injury it inflicted on Bhopal's citizens, the toxic release had other consequences. It poisoned much of the surrounding agricultural land, rendering it unsuitable for either farming or habitation. Medical experts have also observed a significant increase in birth defects, and believe these defects can be traced to toxic exposuure resulting from the accident.

The individual-based compensation scheme does not adequately confront these issues. There need to be additional provisions for detoxifying the soil and assisting former farmers, for addressing the displacement of people who cannot return to their homes, for the medical care of the unborn. Unless these larger problems are addressed, the settlement will not enable the victims of the Bhopal tragedy to rehabilitate themselves and rebuild their lifes. To respond to these problems, the distribution scheme must look beyond individual compensation to the welfare of both the present and future Bhopal community.

Pooling Resources

There is an alternative to the limitations of the individual-based distribution scheme—community-based compensation. Under this model, only a portion of settlement funds are distributed directly to individuals. A significant portion of settlement funds are invested in institutions, programs and services that will collectively benefit the entire class of injured people. The theory behind this alternative distribution scheme is straightforward. The individuals will derive the greatest benefit from pooling a portion of their individual claims, and developing an ongoing financial framework to deal with the Bhopal community's problems and needs.

Under the community model, instead of providing each individual with funds to cover medical costs, the Indian government would invest in modern hospitals to treat the injured. These hospitals would be specifically designed to meet the needs of the Bhopal victims, with staff and facilities to deal with issues such as post-traumatic stress syndrome, genetic counseling and mental health in general. Under the community model, instead of providing individuals with funds to cover soil detoxification and family relocation costs, the Indian government would invest in programs and services designed to address these problems. These programs would target the Bhopal community, providing relief to both current and future generations.

Beyond the immediate problems of medical care, land detoxification and population relocation, the community model could also effectively address other local problems. Settlement funds could be used to establish schools, programs for job retraining and new housing construction. While these expenditures and services may not be directly related to the Bhopal disaster, they would nonetheless help to improve the welfare of the Bhopal community. Because improving the welfare of the Bhopal victims is the central aim of the Union Carbide settlement, these types of programs are wholly appropriate compensation strategies.

The alternative community distribution scheme should not be thought of as completely inconsistent with the individual-based model. It is possible for one compensation scheme to retain elements of both. For instance, the Indian government could offer individual-based compensation for death and permanent, long-term injuries. For all other injuries, people could be given free access to the programs and services established under the community-based model.

The Bhopal settlement provided the Indian government with $470 million to assist and benefit the Bhopal disaster's victims. Having secured these funds, the Indian Supreme Court must now help the compensation courts and local authorities develop a fair and effective distribution scheme. While the existing individual-based compensation scheme is well-intended, it possesses many shortcomings.[4] Because these flaws have rendered the scheme ineffective, the Supreme Court of India should consider alternative compensation models. As Gary Cohen, an advisor to the Bhopal People's Health and Documentation Clinic, commented: "The Bhopal accident highlights the fact that the current legal and medical mechanisms cannot cope with the long-term consequences of toxic industrial disasters where the resulting health and social problems span decades, and even generations."[5]

Recognizing the shortcomings of the current relief scheme, in 1994 the International Medical Commission on Bhopal (IMCB) was created. Although the IMCB's focus is primarily on assessing the health conditions and needs of Bhopal's victims, this health assessment will likely lead to a critical review of how the settlement funds are being spent. As the IMCB turns to the question of allocation, it should consider the advantages of integrating elements of the community-based compensation model into the overall Bhopal strategy. This would help India to better serve the critical medical, social and economic needs of Bhopal's victims.

Chapter 9

Refoliating Vietnam

A Second War for the Forests

Vietnam's forests were devastated by the American War from 1961–72. An August 1994 report by Hanoi's Mangrove Ecosystem Research Centre concluded that U.S. chemical defoliation operations destroyed more than one-third of the mangrove forests in Southern Vietnam's Mekong Delta.[1] The poverty during wartime also led to accelerated destruction of the central and northern forests as citizens became desperate for fuelwood and farmland. All told, the war claimed over 2.2 million hectares of forests.[2] This defoliation and forest destruction left Vietnam with a host of severe health and environmental problems, including Agent Orange contamination, soil erosion, and loss of animal and fish habitat.

Vietnam's forests, however, managed to stage somewhat of a recovery. After the war, Vietnam initiated a massive mangrove reforestation program in the south, and established protected regions for mixed evergreen forests in the central and northern highlands. The initial success of these programs suggested that, unlike many of its Southeast Asian neighbors such as Indonesia and the Philippines, Vietnam might manage to prevent the wholesale destruction of its natural forests. The war may have inflicted great damage on the forests, but this damage was largely a product of the American military, not underlying economic forces. With the war over and the U.S. military gone, the prospects for forest ecosystem recovery looked good.

Unfortunately, these prospects have been dampened by recent developments. Increasingly, population pressures and poverty have begun to place severe demands on the forest.[3] These forces have been encouraged, or at least facilitated, by the Vietnamese government's focus on increasing short-term agricultural productivity. Under current policies, the central government sets regional agricultural goals and local officials are responsible for ensuring that these goals are met. Unfortunately, these economic targets often fail to account for ecological limitations.

As a result of these pressures and policies, Vietnam's forests are once again under siege. Forests are being felled at rates that do not allow for regrowth. According to the United Nations, logging and "slash and burn" agriculture are currently destroying more than 375,00 hectares of natural forest every year in Vietnam.[4]

The best hope to preserve Vietnam's natural forests lies in the legal and political reforms that were initiated in the early 1990s. While the primary aim of these changes has been to introduce market elements into the Vietnamese economy, these reforms have also granted additional free speech rights and attempted to limit government abuse by bringing administrative activities under the rule of law.[5] As such, the reforms closely resemble the policy of *perestroika* launched in the former Soviet Union in the mid and late 1980s. As with *perestroika*, deteriorating environmental conditions have served as a catalyst to, and a focal point for, legal reform efforts. It remains to be seen, however, whether these reforms will be deep enough and come in time to protect Vietnam's threatened forests.

A Period of Reform

After the defeat of South Vietnam and national unification in 1975, Vietnam operated a centrally-planned economy controlled by the Communist Party. This political system was extremely authoritarian, and provided little room for public debate or government criticism. It also created an economic atmosphere that discouraged citizen initiative as well as foreign investment. As a result, Vietnam's standard of living declined dramatically after the war. By the mid-1980s average annual per capita income was under $200.

In response to these economic conditions, Vietnam began experimenting with "mixed market" politics, or *Doi Moi* (renovation) in the early 1990s.[6] The *Doi Moi* reforms were initially intended to be strictly economic, and were not intended to affect Vietnam's basic political system or reduce the broad powers of the Communist Party. As with *perstroika* in the former U.S.S.R., however, economic reform in Vietnam eventually led to political debate over other issues. The weak status of the rule of law and the declining condition of Vietnam's environment were among the issues that were later included in the *Doi Moi* program.

The legal system has undergone several important changes during this period of reform. First, prior to *Doi Moi*, Vietnamese legal rights were viewed primarily in terms of citizens' duties to the state and the party. Laws were meant to circumscribe the behavior and liberties of individuals, not the government. *Doi Moi*, however, introduced reforms that set forth obligations and prohibitions concerning the actions of the state. These should help bring state activity under the rule of law, and make government officials more accountable.

Prior to *Doi Moi*, assertions of illegal behavior could only be presented to People Committees, who then decided whether to initiate administrative action on the complaining party's behalf. Although there is still no formal judiciary, Vietnam has now created special economic courts to help resolve disputes between foreign and domestic companies officially registered with the state. It is expected that the jurisdiction of these special economic courts will be expanded, and that they may soon provide a forum for other disputes and a broader range of claimants. While this forum would lack the independence of a separate judiciary, it would nonetheless represent an important step toward establishing the rule of law.

Significant changes have also been implemented in the environmental field. First, in 1991 Vietnam developed the National Plan for Environment and Sustainable Development (NPESD), which sets forth the government's environmental policy priorities.[7] The NPESD calls for expanding the system of nature preserves, increasing Vietnam's participation in international conservation programs, and greater funding for reforestation efforts.

Second, in 1991, Vietnam adopted the Law on Forest Protection and Development (Forest Law) and in 1994 it enacted the comprehensive Law on Environment. The Forest Law requires that the Vietnamese government preserve the forest resources of the nation, restore barren lands within the forestry sector, and protect watersheds and threatened wildlife. The Law on Environment calls upon state agencies to establish plans to prevent environmental degradation and prohibits activities that "destroy ecological equilibrium." It also provided that persons who take advantage of their position or power to infringe environmental protection laws shall be disciplined or criminally prosecuted.

Third, in 1992 a new comprehensive environmental agency was created, the Ministry of Science, Technology and Environment (MOSTE). MOSTE's responsibilities include the environmental review of projects proposed by other agencies, such as Ministry of Agriculture and Food Industries and the State Commission on Corporate Investment (SCCI). MOSTE's review and approval is an administrative prereuqisite to final project authorization.

Lastly, in 1993 Vietnam signed the United Nations Biodiversity Convention. Under the terms of this international treaty, nations must "promote the protection of ecosystems, natural habitats and the maintenance of viable populations of species in natural surroundings." The primary aim of the Biodiversity Convention is to encourage national governments to increase their domestic commitment to ecosystem preservation and wildlife protection. The Convention is also expected to serve as a major financing mechanism for global biodiversity conservation. By signing the Convention, Vietnam might therefore be able to secure international funds to assist in the development and implementation of forest conservation programs.

Taken together, these legal and environmental reforms have the potential to profoundly and positively impact forest management in Vietnam. Translating this potential into reality, however, has not been easy. Significant economic and social hurdles have hindered attempts to preserve Vietnam's forests.

The Pressures of Poverty

The far-reaching legal reforms in the environmental fields suggest that, in terms of forest protection, Vietnam has set itself on the right course. The successful implementation of these changes would provide a solid legal framework to ensure the sustainable management of the nation's forests. The process of implementation, however, has been difficult and slow. Rural poverty and population pressures have impeded forest protection efforts, and have allowed deforestation to continue.

Since the end of the American War in 1975, Vietnam's population has increased to over seventy million people. This population is clustered around two major deltas—the Mekong River Delta in the south and the Red River Delta in the north. The population increase has placed tremendous environmental stress on rural areas, particularly in the two deltas where the vast majority of Vietnamese live.[8] The forests, in particular, have suffered greatly. Poverty and dire economic conditions have to led to over-exploitation of forest resources.

In southern Vietnam's Mekong Delta, mangrove forests are being rapidly cut down for firewood and to clear land for agriculture and aquaculture (fish and shrimp ponds).[9] This logging and forest conversion has been carried out in an indiscriminate and environmentally unsustainable manner. Erosion has caused river siltation and destroyed the biological integrity of many streams. Boars, monkeys, foxes, boas and migratory birds such as cranes have lost important habitat and breeding grounds.[10] Moreover, most of the aquaculture ponds are poorly designed and have failed to provide adequate fresh water flow. As a result, ponds generally remain productive for two or three years and are then abandoned. The economic benefit derived from mangrove deforestation is therefore minimal and short-lived. Although the government has not endorsed the logging and conversion of mangrove forests, it has so far done little to stop the destructive cutting.

Similar problems and pressures confront forests in other regions of Vietnam. Upstream from the Mekong Delta, in An Giang and Dong Thap provinces near the Cambodian border, hillside forests are being cleared for agricultural land. The removal of hillside trees has left slopes exposed and unprotected. This loss of water catchment has resulted in increased run-off during heavy rains. This, in turn, has led not only to soil erosion but to destructive floods downstream. According to environmental officials in Can Tho, these floods have left hundreds of people dead and devastated the local farming economy.

The evergreen forests in the central Vietnam's Dalat Plateau have also been targeted for intensive logging. It is estimated that the forest cover in Vietnam's highlands has been cut to fifty-six percent.[11] As with the cutting in An Giang and Dong Thap, much of the logging in Dalat has taken place on steep slopes. While some of this logging has been for firewood and farmland, there are also reports of raw logs being smuggled to the coast for foreign export. To ensure that Vietnam captures the full value of finished wood products, the government recently banned the export of unprocessed logs. Local authorities, however, apparently lack the resources to effectively enforce the export ban. Many environmentalists also believe that provincial officials are being paid off to ignore these export restrictions.

Mountainside deforestation in the Dalat Plateau has washed away large amounts of soil, damaging crops and farmland below. It has also greatly diminished the ecological richness and scenic qualities of what was once considered one of Vietnam's most beautiful and biologically diverse wilderness areas.

On a national scale, the environmental impact of deforestation has been staggering. An alarming forty percent of Vietnam, is now classified as bare land.[12] Although about one million of this is accounted for by rocky mountains. The rest is land that was formerly forests but that has been cleared and degraded to a condition where it is no longer biologically productive.

Additionally, a 1991 report by the government Forest Inventory and Planning Institute (FIPI) concluded that logging was destroying the critical habitat of moose and wild goats, and was pushing populations of skid deer into extinction. The FIPI report warned that because there has been "inadequate attention to take care of and maintain the environment for wild resources to regenerate, many species' existence is now problematic and even endangered."[13]

The Fate of Vietnam's Forests

The degraded condition of Vietnam's forests reveal the current disparity between official forest policy and actual forest protection. Some of this disparity can perhaps be attributed to corruption and

Vietnam's continuing reliance on outdated forest management models. Much of it, however, is a reflection of Vietnam's larger struggle to bring state policy under the rule of law. In this regard, environmental and forestry law are encountering the same basic obstacles as the banking and commercial fields—namely, the difficulty in developing legal mechanisms that allow citizens to ensure that laws are enforced.

These mechanisms are not currently in place. For instance, under the new Law on Environment, the SCCI cannot authorize foreign economic enterprises until the enterprise proposal has been reviewed and approved by the MOSTE. The SCCI, however, recognizing the weak political power of the new environmental agency, has routinely ignored the MOSTE approval practice. The citizens who will bear the health and environmental brunt of these enterprises currently have few avenues of recourse against the SCCI or the private foreign companies. At present, compliance with environmental laws can only be ensured by applying political pressure at the administrative level. In practice, such recourse means very little.

Forest protection in Vietnam faces a difficult road ahead. This road, however, does point in a general direction. Vietnamese citizens concerned about, and threatened by, the impacts of deforestation need to focus on deepening the legal and environmental reforms that have begun. The jurisdiction of the newly created special economic courts need to be expanded. The political credibility and independence of MOSTE needs to be strengthened. In the short-run, these are the surest means to protect the common and collective ecological heritage of Vietnam's forests. Without the effective means of ensuring that environmental laws are in fact implemented and observed, Vietnam's forests face a bleak future.

Part III.
TRADE'S HARVEST

The modern framework for trade rules and the global economy was established in the late 1940s, with the creation of the General Agreement on Tariffs and Trade (GATT), the International Monetary Fund (IMF), and the World Bank. These institutions were developed in response to two particular historical events—the Great Depression and World War II. At their inception, GATT, IMF, and the World Bank were viewed as a means to stabilize global markets and political conditions, to prevent a reoccurence of the tariff wars, widespread inflation, and currency devaluation that many believed contributed to both economic decline and the rise of fascism.

Since the 1940s trade framework was established, however, two important changes have taken place. First, capital now moves freely across national borders, and thus corporate investment and profits are no longer tied to the national economy. Second, we now have a greater scientific understanding of the extent to which economic activities are degrading the planet's nonrenewable natural resources and basic ecological systems. GATT, IMF and the World Bank were not designed to handle these new developments, and no effective international institutions currently exist that do. The result of this governance vacuum has been an increase in the power and political autonomy of multinational corporations, and a corresponding increase in global environmental degradation.

The current predicament was expressed artfully by Herman Daly, a former Senior Economist with the World Bank and now one the leading advocates for trade reform. Daly asks us to think of the international economy in anatomical terms. GATT, IMF and the World Bank are designed to maintain the circulatory system, to ensure that trade restrictions and currency fluctuations do not block the flow of goods, services and investment—the economic lifeblood. However, the circulatory system is ultimately dependent on the respiratory and digestive systems, which ensure that the lifeblood receives and processes the nutrients it needs to sustain the body. In the global economy, our respiratory and digestive systems—our nonrenewable resources and ecosystems—are in bad condition, and their deterioration has placed our future at great risk. This deterioration will only be halted when the global economy is understood as a component of, not something distinct from, the global ecosystem. International law will play a critical role in establishing and enforcing this new ecological conception of trade.

Chapter 10

A Difficult Swim

The Sea Turtle Navigates GATT

NAFTA and GATT have spawned fierce political debates between promoters and critics of global free trade. One of the most intense debates surrounding these agreements focuses on the impact of NAFTA and GATT trade rules on environmental protection. Many environmentalists have argued that free trade rules, and the international organizations that implement these rules, could undermine U.S. policies that protect the environment and promote conservation. The current row over the impact of shrimp fishing on sea turtles indicates that environmentalists' concerns may be justified.

In April 1996, the United States Court of International Trade (CIT) issued a landmark decision in *Earth Island Institute v. Christopher.*[1] In this case, the CIT ordered the U.S. State, Commerce and Treasury Departments to block the importation of shrimp from all nations that had not adopted adequate policies to protect sea turtles. Worldwide, hundreds of thousands of sea turtles are killed each year as a result of shrimp-harvesting operations, in which the turtles drown trying to escape the shrimp nets.[2]

The CIT based its ruling on an interpretation of a 1989

amendment to the federal Endangered Species Act (ESA), Section 609.[3] Section 609 calls for the development of a shrimp "certification" program by the U.S. federal government. Under this program, nations desiring to export shrimp to the U.S. must be certified by the U.S. government. The U.S. government can only provide this certification if the exporting nation can demonstrate that it catches shrimp using methods that provided a level of protection to sea turtles comparable to protection provided for under U.S. conservation laws.

Although hailed as a victory by marine conservation advocates, the CIT's ruling in *Earth Island* is now under attack. Foreign countries subject to the certification requirements have filed a formal complaint with the World Trade Organization (WTO) alleging that Section 609 is inconsistent with the United State's trade obligations.[4] The WTO is the successor organization to GATT, and is now charged with overseeing the agreement's implementation.

The nations bringing the WTO challenge allege that the U.S. sea turtle protection program violates GATT rules that prohibit trade restrictions based on extraterritorial conservation goals and the methods by which products are produced or harvested. If this challenge proves successful, the United States could be subject to countervailing import restrictions, as well as powerful diplomatic pressure to bring its policies into compliance with GATT. This, in turn, could lead the United States to weaken or block implementation of the turtle protection program—the very scenario forecast by environmentalists.

The Sea Turtle Litigation

The population of sea turtles worldwide is threatened by destructive fishing practices.[5] It is estimated that over 125,000 turtles die every year, not to serve as food for people, but because they are hauled in (and drowned) as unwanted bycatch for target catch such as shrimp and tuna. According to the National Marine Fisheries Service (NMFS), a U.S. Commerce Department subagency responsible for implementing the federal Endangered Species Act, there are currently at least four species of sea turtles that now face possible extinction: the loggerhead, the green leatherback, the hawksbill and Kemp's Ridley.[6] Of these species, the Kemp's Ridley turtle is the most threatened, with less than 1,500 nesting turtles remaining in the wild.[7]

In response to these threats, the U.S. Congress has adopted policies to reduce the destructive impact of fishing on sea turtles. In terms of shrimp trawlers, U.S. regulation has focused primarily on mandating or encouraging the use of turtle-exclusion devices (TEDs). TEDs are metal trap-doors attached to shrimp nets that enable turtles to escape nets and thereby escape drowning. It is estimated that TED's can reduce sea turtle mortality from shrimp fishing operations by 97 percent. In 1987, the NMFS promulgated regulations under the federal Endangered Species Act (ESA) requiring that the entire U.S. shrimp fishing fleet use TEDs. In 1989, Congress adopted Section 609 of the ESA, creating the certification program for nations desiring to import shrimp into the United States. Section 609 requires the secretaries of State, Commerce, and Treasury to prohibit the importation of shrimp products from all countries that have failed to mandate shrimp fishing practices that provide sea turtle protection comparable to that provided under U.S. law. The use of TEDs is the primary means to ensure comparable levels of sea turtle protection

The economic reasoning behind, and the policy objectives of, Section 609 were straightforward. The United States is a major purchaser of shrimp. According to a 1995 U.S. Commerce Department report, U.S. shrimp imports total more than $1.2 billion annually.[8] Foreign nations that export shrimp have a clear interest in being able to sell their product to the lucrative U.S. market. By conditioning access to the U.S. market on adherence to minimal sea turtle protection standards, foreign nations are encouraged to improve the environmental performance of their shrimping industry. Moreover, the import requirement creates a level playing field for U.S. fishermen, who must already comply with the 1987 turtle protection regulations.

Although the language of Section 609 stated that the sea turtle protection requirements applied to all countries, the State Department issued regulations that limited the geographic scope of this language. These regulations interpreted Section 609 as applying only to shrimp fishing nations in the Atlantic/Caribbean region.[9] The State Department justified this limited interpretation of Section 609 on the grounds that Congress only intended the TED requirement to apply to sea turtles that are caught in, or migrate through, U.S. coastal waters.

The State Department's decision to limit Section 609's application to Atlantic/Caribbean shrimp fishing nations was intensely criticized by marine conservation advocates. The most prominent of these critics was Earth Island Institute (Earth Island), a non-governmental organization based in San Francisco that directs the Sea Turtle Restoration Project. In 1995, Earth Island filed suit against the Departments of State, Commerce and Treasury in the U.S. Court of International Trade (CIT). Earth Island challenged the State Department regulations limiting the application of Section 609, and asked the court to order the State Department, as well as other federal agencies to apply the shrimp certification program to all foreign countries, regardless of geographic location.

The defendants responded to Earth Island's claim with two basic arguments. First, they argued that Earth Island lacked standing to bring the suit, because the group did have a legally sufficient interest in ensuring Section 609's full implementation. Second, the government argued that the State Department's regulation limiting the geographic scope of Section 609 was a reasonable interpretation of the underlying legislation.

The CIT rejected both of the defendants' arguments. The court issued an order compelling the federal government to prohibit the importation of shrimp or shrimp products into the United States from all foreign nations that have not reduced sea turtle mortality from shrimp fishing operations by ninety-seven percent, the level that can be achieved through the use of TEDs. Furthermore, the CIT ordered the government to comply with the order no later than May 1, 1996.

In an effort to limit the scope of the CIT's April 1996 ruling, the State Department promulgated new regulations to implement the shrimp certification program. These regulations held that a foreign country would be certified so long as the particular shrimp being imported into the United States were caught using turtle-safe methods.[10] Under the regulations, foreign nations did not have to demonstrate that their entire shrimping fleet was turtle-safe to be certified. As such, the regulations attempted to once again limit the scope of Section 609.

In response, Earth Island filed a motion with the CIT contending that the State Department regulations were not in conformity with Section 609 and the April 1996 CIT ruling.[11] According the Earth Island, the State Department regulations "eviscerated" Congress' purposes because foreign countries can "evade the law's embargo by exporting to the United State those shrimp caught by a few designated vessels which are equipped with TEDs, while exporting elsewhere shrimp caught by those which are not." The CIT agreed with Earth Island, and issued a strongly worded opinion against the Clinton Administration.

The Turtle–GATT Clash

In addition to arguments based on standing and the reasonableness of the State Department's regulations, the defendants in the *Earth Island* litigation also offered a third argument in support of their limited geographic application of Section 609. This third argument focused on the United States' international free trade obligations under the GATT. According to the government, the shrimp certification program established under Section 609 conflicted with GATT's trade rules, and federal agencies, as well as the CIT, should interpret Section 609 to minimize or avoid these conflicts.

The defendant's free trade argument was based, at least in part, on the rulings and interpretations of GATT dispute panels. These panels are convened to help resolve conflicts among GATT's contracting parties. Several of these panels have considered the relation of environmental trade measures to GATT's free trade requirements. The two most significant panel rulings concerned a Canadian restriction on the export of unprocessed salmon and herring,[12] and a United States' restriction on the import of tuna caught in purse-seines nets.[13]

In the Canadian fish processing dispute, Canada argued that its processing requirement was justified because it closely related to the country's efforts to prevent over-fishing by ensuring that Canadian fishermen did not lose the value of their more limited catch. The GATT disagreed, concluding that the processing requirement was not "necessary" to protect fish stocks, and that its primary purpose was to protect the domestic Canadian fishing industry, not to promote conservation.

In the tuna purse-seine net controversy, the GATT considered a U.S. law banning the import of tuna caught in a manner that results in a high dolphin mortality rate. The United States offered two arguments in defense of its import restriction. First, it argued that the restriction was non-discriminatory because the ban on tuna caught in purse-seines nets applied to both domestic and foreign parties. Second, the United States claimed that, even if the import restriction was found to discriminate against certain foreign nations, such discrimination was justified under GATT's conservation exceptions. These exceptions permit trade restrictions "necessary to protect human, animal, or plant life and health," and measures "relating to the conservation of exhaustible natural resources."

The GATT rejected both of these arguments. First, it found that the import restriction applied to production methods (fishing practices), not products, and that GATT did not permit production-based trade restrictions. Second, the panel found that because the restriction was "primarily aimed" at forcing other countries to change their policies, not marine conservation, it did not fall within the meaning of GATT's environmental and conservation exceptions.

In the *Earth Island* litigation, these rulings provided the legal background for the defendants' contention that Section 609 was inconsistent with the United State's GATT obligations. The government argued that broad application of Section 609 would likely prompt a formal GATT challenge, and that, under principles established in previous panel rulings, the U.S. was likely to lose this challenge. In considering this argument, the CIT conceded that it was appropriate to "seek to minimize or reduce conflict to the maximum extent possible...consistent with Section 609's basic statutory purposes." It went on to conclude, however, that "the record of enforcement of Section 609 to date does not reveal troubling tensions with the foreign sovereigns already deemed covered, including those not certified positively and subject to embargoes." As such, the CIT seemed to suggest that Section 609's alleged conflict with GATT was too speculative to warrant limiting the application of the shrimp certification program.

Although not addressed directly by the CIT, there may be an additional reason for rejecting the government's argument concerning the impact of U.S. GATT obligations. Increasingly, trade and environmental experts have begun to question the legal grounds supporting the distinction between measures regulating products, and measures regulating production methods. More specifically, there is now considerable support for the position that GATT's non-discrimination rules relating to products have been fundamentally misunderstood, and misapplied, by GATT dispute panels.[14] Steve Charnovitz, Director of Yale's Global Environment and Trade Study, has been one of strongest voices calling for this reassessment.[15]

Charnovitz maintains that, at the time GATT was adopted in 1947, there were already many longstanding examples of trade restrictions based on production methods relating to environmental concerns. Although GATT dispute panels, and many trade experts, have assumed that the 1947 agreement sought to outlaw these existing measures, Charnovitz finds no evidence to support this assumption. Conversely, he maintains that GATT actually reaffirmed the appropriateness of such production-based measures, and that a "good case can be made that the GATT has green roots."[16]

Aside from the issue of whether GATT envisioned process-based trade restrictions when it was adopted in 1947, there are persuasive arguments to support the contention that such an interpretation is nonetheless appropriate today. In terms of ecology, we now understand, much clearer than we did in 1947, how the manner is which resources are extracted and processesd contributes to environmental damage. For instance, the most significant damage caused by logging of natural forests is the loss of biodiversity, species habitat and carbon sinks, not the environmental attributes of wood or paper. Similarly, the most significant damage caused by unsustainable industrial agriculture is topsoil loss, desertification and water pollution, not poor food quality. Given our current understanding of ecology, the distinction between products and production methods makes little sense. Instead of clinging to an arbitrary and outdated distinction, we must develop new trade rules that are consistent with with our modern environmental knowledge.

While the CIT did not find it necessary to reconsider the product/production method distinction in *Earth Island v. Christopher*, the WTO may soon have such an opportunity. In response to the CIT's April 1996 decision, many nations, including all six members of the Association of Southeast Asian Counties (ASEAN) threatened to file a formal challenge with the WTO. In July 1996,

Suvit Khunkitti, Thailand's Agricultural Minister, warned that unless the United States eased the ban, ASEAN members would raise the issue at the World Trade Organization's December 1996 meeting in Singapore.[17]

When the Clinton Administration declared that it would comply with the CIT's ruling, Thailand made good on its threat. In December 1996, Thailand, joined by Malaysia and Pakistan, filed a formal complaint with the WTO. In April 1997, the WTO appointed a dispute panel to rule on the complaint. As of Janaury 1998, the parties had submitted written arguments and the WTO had organized a special scientific panel to help resolve conflicting factual claims. No final decision had been released.

Now that a formal challenge has been brought before the WTO regarding the U.S. shrimp certification program, the CIT's resolution of the GATT-consistency issue may prove inadequate. With a formal WTO challenge filed, the "troubling tensions" that were absent in the CIT case may now be present, and the issue of GATT-consistency no longer appears to be a matter of mere legal speculation. With the convening of the WTO dispute panel, several difficult questions have arisen. Should the WTO reject the product/production method distinction recognized in earlier panel decisions? If the WTO finds Section 609 inconsistent with international trade rules, should the U.S. Congress amend the law, or should the U.S. courts reevaluate the CIT's April 1996 decision? The fate of the U.S. sea turtle protection program will hinge on the answers to these questions.

Swimming Beneath the Surface

The sea turtle faces an uncertain future. Although the CIT's *Earth Island* decision has helped elevate the legal status of marine conservation, this victory may be short-lived. In addition to the WTO challenge, there have been calls for Congress to amend

Section 609.[18] Furthermore, the decision has also been appealed. These efforts threaten to weaken, and perhaps overturn, the CIT's landmark decision.

At least for now, however, the sea turtle has managed to achieve some progress. The key element to this progress is the recognition that, under U.S. law, species should not be driven into the extinction by destructive and outdated fishing practices. Whether this right to exist will hold up under international trade rules remains to be seen. One thing is certain, however. Despite the technical language that will likely dominate the turtle-trade debate, this right swims just beneath the surface.

Chapter 11

Trees Falling
Forests and the Timber Trade

The term *free trade* suggests a global economy in which governments do not interfere with private business transactions and in which the market naturally discourages activities that are not economically beneficial or efficient. The above definitions do not reflect current reality. Particularly in the international timber trade, governments continue to protect and promote logging interests, and the international market does not discourage logging practices that result in widespread and long-term economic damage.

The international timber trade is one of the most significant economic forces contributing to the unsustainable logging of natural forests. Although pressures to clear natural forests for farmland, rangeland and fuelwood play some role in this destruction, a comprehensive 1995 book by the World Wildlife Fund (WWF) concluded that these forces are of secondary economic significance. The book, *Bad Harvest: The International Timber Trade and Degradation of the World's Forests*, was based on more than fifteen years of research by Nigel Dudley, Jean-Paul Jeanrenaud and Francis Sullivan, three leading international forestry experts. In it, WWF found that "far from being a negligible cause, the timber trade is the primary cause of forest degradation and loss in many of the remaining natural forests."

The timber and wood products trade, however, is not shaped by the invisible hand of supply and demand. It is shaped by governments and corporations whose goal is to keep the demand for wood products high and supply costs (such as environmental protection) low. This is why the Pacific Rim has lost fifty percent of its native forests in the last twenty-five years and why the countries which have lost the most forest, such as Indonesia and the Philippines, have sunk even deeper into poverty.[1]

Because of the transnational nature of the timber trade, local and national forest protection efforts have by and large failed to halt deforestation on a global scale. Protecting one region's forest has often simply resulted in the destruction of another region's forests. To prevent this transfer of harm, forest protection advocates are now looking beyond piecemeal conservation strategies. They are looking at ways to subvert the timber trade paradigm that has placed the world's forest at risk.

The Pacific Rim Cut

The forests of the earth, and of the Pacific Rim in particular, are being logged at an ecologically unsustainable pace. The best available evidence indicates that nearly one-half of the Pacific Rim's forests have been destroyed in the last half century. The environmental and social consequences of this deforestation have been profound. Thousands of species have been forced into extinction.[2] Soil erosion and watershed degradation have rendered river waters undrinkable for humans and unlivable for fish. Indigenous populations in Brazil, Canada and Southeast Asia have been forcibly evicted from their traditional lands.

In the United States, the Spotted Owl litigation and the Pacific Northwest logging disputes have focused the American public's attention on the plight of our national forests. Less attention, however, has been given to explaining how this regional controversy fits into the global picture. More specifically, inadequate attention has been focused on how the international timber trade is contributing to forest destruction both at home and abroad.

In Siberia, for example, coniferous forests cover an area equal in size to the continental United States. These forests, which serve as critical habitat for the endangered Siberian tiger and Amur leopard, have been targeted for intensive logging by Japanese, Korean and U.S. timber companies.[3] These companies have been attracted to the region by the enormous tracts of mature forests, and the absence of forest protection obstacles found elsewhere.

In late 1994, a coalition of mill operators in the U.S. Pacific Northwest called Global Forestry Management Group announced plans to log and export 100 million board feet a year from Siberia.[4] These unprocessed logs were intended to supply mills in the U.S. where almost all private old-growth had been logged and most of the remaining old growth on federal lands was protected. Global Forestry Management Group's plan was greatly facilitated by a June 1994 Memorandum of Understanding between U.S. Vice-President Al Gore and Russian Prime Minister Viktor Chernomyrdin.[5] The stated goal of the Gore-Chernomyrdin agreement was to promote cooperation in the forest products industries. Given the absence of effective forest protection policies in Siberia and the Russian Far East, however, the probable result will be increased forest destruction.

Several federal U.S. agencies have also played a role in promoting the logging and export of Siberian forests. The U.S. Trade and Development Agency (TDA) spent $500,000 on an economic feasability study for the Global Forestry Management Group plan. Similarly, another federal agency, the Overseas Private Investment Corporation (OPIC), is providing risk insurance to U.S. timber companies logging in Siberia. While the U.S. government claims that TDA and OPIC are merely facilitating economic cooperation, others contend that the agency financing amounts to a subsidy. It reduces the risk and helps lay the political groundwork for logging operations abroad.

The story is similar in Chile, where the forest sector is also driven by international demand and government support. From 1973 to 1990, wood product exports in Chile increased from $39 million to $760 million. During this same period, the national government extended over $88 million in subsidies to domestic logging companies.[6]

Much of the increase in Chilean exports has been fueled by the Japanese market, where chips and pulp are needed to make disposable paper products. Much of the wood exported to Japan comes directly from Chile's natural forests. Some also comes from managed tree farms. These tree farms, however, are often on land where native forests have recently been logged out. As Manfred Max-Need, 1993 Chilean presidential candidate observed: "There are mountains of woodchips waiting to be sent to Japan to be made into toilet paper. I don't think that is a very noble destiny for our native forests."[7]

The Siberian and Chilean situations are regional parts of a global pattern. To satisfy the worldwide demand for wood, the logging industry seeks the path of least resistance, logging in regions where environmental restrictions are minimal and government subsidies are plentiful. As such, the international timber trade is truly a moveable feast—one that has placed forests, wildlife and indigenous people at risk.

The Timber Trade Paradigm

Although demand may help fuel the engine of deforestation, this demand is not simply the product of the unfettered marketplace. Rather, it is the intended result of political and institutional arrangements, operating at both the national and international levels. These arrangements are designed to benefit identifiable interests, namely the timber, wood, paper and pulp industries.

At the national level, many governments continue to grant logging contracts on public lands with little or no stumpage fees. These concessionary rates do not incorporate the full costs of environmental and economic damage associated with logging and do not incorporate the full range of environmental and economic benefits provided by forest protection. Moreover, national governments continue to underwrite road construction and other infrastructure costs associated with large-scale logging activities. These arrangements lower the production costs for the big timber and wood companies and thereby make destructive logging more profitable. These arrangements also undercut the competitiveness of smaller, locally-based timber companies that are committed to practicing ecologically sustainable forestry.

Nationally, there is also a recurring pattern of collusion between timber interests and the environmental agencies charged with regulating these industries. The ability of industries to control the policies and enforcement of regulating agencies is sometimes called *agency capture*. Sometimes agency capture is achieved under color of law, such as when Chile's Pinochet routinely handed out logging concessions to timber companies that supported his dictatorship. Other times, agency capture is achieved through illegal means, such as when the Malaysian forest minister reportedly received several million dollars from timber industry connections. Whether legal or illegal, the results of agency capture are the same—regulatory agencies fail to implement, or ignore, environmental protection standards that would decrease the short-term profitability of logging.

Similar destructive arrangements can be found at the international level. The General Agreement on Tariffs and Trade (GATT) and the North American Free Trade Agreement (NAFTA), for instance, forbid import restrictions that discriminate on the basis of production or harvesting methods. The method of logging, and the forest protection policies associated with logging, are examples of such harvesting methods. The ban on production-based import rules has already impacted one national effort to adopt sustainable forestry trade provisions. In 1992, Austria adopted a mandatory labeling scheme for all imported timber in which nations would have to demonstrate that the timber was sustainably harvested.[8] Several nations asserted that Austria's labeling scheme constituted an impermissible trade barrier under GATT. Fearing that GATT might impose sanctions or authorize countervailing trade measures, Austria repealed the law.

A labeling regime similar to the repealed Austrian law is being developed by the European Union. The proposed EU regime would create an "Eco-Label" to designate products (not only timber) that meet certain objective environmental standards. Unlike the Austria law, the purpose of the Eco-label is not to regulate the importation of products. Rather, its purpose is to provide European consumers with information about the environmental characteristics of products, and to provide an incentive for producers, both in Europe and abroad, to maintain high environmental standards.[9]

When the EU announced in 1995 that it intented to apply the regime to wood products, Jack Creighton, chief executive of timber multinational corporation Weyerhauser attacked the proposal as an unacceptable trade barrier. Creighton maintained that the proposal was merely a "pretext to keep American and Canadian forests products out of Europe," and that the scheme could provoke a formal GATT challenge.[10]

The funding policies of international lending institutions, such as the World Bank and the International Monetary Fund (IMF), have also contributed to global deforestation. Since its creation after World War II, the World Bank has provided billions of dollars to further economic development in the Third World. Most of these development loans were authorized without environmental impact assessment, and without the public participation of affected citizens. Just as with domestic subsidies, these international funds have helped underwrite the costs of logging and stimulated wood exports from the developing world.

The World Bank's colonization project in Polonoreste, Brazil, in the mid 1980s is an excellent illustration of the nexus between international lending and deforestation.[11] In Polonoreste, the World Bank financed a project to create new rural settlements and promote subsistence agriculture. The project, which was intended to reduce urban population pressures, called for the construction of a 1,500 kilometer paved highway through the heart of the Amazon basin. The results of the project were economically and environmentally disastrous. The highway construction, as well as slash and burn land clearing for agriculture (made possible by access from the new highway), led to widespread deforestation. Forests containing endangered and irreplaceable biodiversity were destroyed. In addition, thousands of forest-dwelling indigenous people were forced off their traditional lands.

In response to widespread criticism of projects like the one in Polonoreste, in 1991 the World Bank developed new environmental guidelines. These guidelines require the preparation of an environmental assessment for all World Bank projects.[12] Although these responses are a step in the right direction, so far they have not significantly improved the World Bank's environmental performance. While the new environmental assessment process often identifies environmental problems, these problems are usually ignored or downplayed at the loan approval stage. As Bruce Rich, International Program Director for the Environmental Defense Fund, explains: "The bank's regional environmental staff are supposed to exercise close scrutiny over projects, but hampered by both limited budgets and limited authority, they are all but powerless to stop ambitious country directors from riding roughshod over bank policies."[13]

Unfortunately, many assistance agencies in the Japan and Europe are following the World Bank's example. Japan's Overseas Development Agency (ODA) and the Swedish Development

Agency (SDA) have also funded and underwritten forestry projects in developing countries without adequate assessment of the environmental consequences.[14] For instance Japan's ODA has helped subsidize forest road building in Burma, Indonesia, Thailand and other Southeast Asian countries. The official justification for the ODA's projects was that local people want to use the roads. In most cases, however, the logging roads were cut where they were useful for loggers, not the local people, and the roads were generally not maintained after logging was finished. Similarly, in 1993 the Swedish Development Agency helped finance a huge pulp and paper mill in Vietnam. This mill relies on over 70,000 hectares of natural bamboo, and has been heavily criticized by environmentalists for its impact on Vietnam's natural vegetation and ecosystems. Instead of learning from the tragic World Bank experience, Japan and Sweden's development agencies appear to be repeating it.

The IMF's role in promoting deforestation has been less direct than that of the World Bank, but no less significant.[15] In the early 1980s, the IMF developed a funding strategy in response to the international debt crisis. Under this strategy, IMF provided funds to developing countries to help with debt payments. IMF funds, however, were tied to conditionalities. These conditionalities, or structural readjustments, required debtor countries to sharply reduce public spending (austerity) and to adopt trade policies that were more conducive to foreign investment.

In terms of forests, the IMF conditionalities have had several unfortunate results. First, under an imposed policy of austerity, governments reduced funding for environmental planning, protection and enforcement. This led to increased mismanagement of forests and illegal logging. Second, the foreign investment that followed the IMF's trade and economic reforms tended to be in resource intensive sectors such as logging, mining and oil exploration. This also placed new pressures on the forest. As such, the structural adjustments demanded by the IMF reinforced and encouraged destructive logging in the developing world.

Reconstructing the Marketplace

The national and international arrangements that drive the timber trade are not the product of mysterious market forces. They are the work of identifiable parties furthering their economic interests through political influence. Until this framework is dismantled, the timber trade will continue to threaten and destroy the world's forests. This dismantling requires removing the institutional incentives that encourage unsustainable logging.

Forest protection advocates have already developed strategies for promoting sustainable forestry at the national level. These strategies include adopting legislation that expressly protects species and ecosystems, defending the land tenure rights of indigenous people, requiring environmental impact assessment and citizen participation, ending direct and indirect subsidies to the timber industry, promoting the use of alternatives to wood-based products, and certification of ecologically sustainable timber sources. Although these strategies are currently under attack in the United States and abroad, they nonetheless provide a blueprint for national forest protection efforts. They provide a set of coherent and tested policies around which forest protection advocates can focus their efforts.

Despite the significant number of international institutions involved in forestry issues, such as the International Tropical Timber Association (ITTA), the United Nations Food and Agricultural Organization (FAO), the Commission on Sustainable Development (CSD), the Global Environmental Facility (GEF) and the World Trade Organization's (WTO) Committee on Trade and the Environment, this blueprint is currently lacking at the international level. These institutions have yet to effectively address and respond to the role that investment, consumption, monetary policy, debt and trade rules play in deforestation.

Arguably, it is the sheer number of international institutions involved in forestry issues that has made progress so difficult.

Nations and economic interests opposed to increased forest conservation have so far managed to play these different institutions off each other.

Whenever serious reforms, such as changing consumption, debt restructuring, or changing trade and investment rules have been proposed in one institution, these nations and economic interests have maintained that a different institution is the appropriate forum to discuss and resolve such issues. So far this strategy has worked, leaving forest protection advocates both exhausted and ineffectual.

This problem was analyzed by David Humphreys in his 1996 book *Forest Politics: The Evolution of International Cooperation.* Humphreys views this institution-shifting technique as "an attempt to deflect unwelcome attention on an issue in one forum into another, presumably with the intention that the latter will not have the time or the adequate mandate to deal meaningfully with the issue in question."[16] Environmentalists have yet to develop an effective response to this tactic, and thus have been unable to control the debate over deforestation and the timber trade.

The time is now ripe for forest protection advocates to critically evaluate the effectiveness of, and perhaps more importantly, the relationship among, the ITTO, FAO, CSD, GEF and WTO, and to develop a coherent international agenda. The focus of this agenda should be strengthening national forest policies, on ensuring that corporations and nations do not export the economic, ecological and human rights consequences of unsustainable logging to countries with lower environmental standards. This will require changing the existing rules of trade so that, at least in the case of timber and ecosystem management, national import and export laws can actively discourage destructive logging and actively promote responsible forestry. It will also require a critical examination of the ways international debt and development projects are driving the economics of deforestation.

In short, to curtail the destructive tendencies of the timber trade, the rules governing the import and export of forest-based products will have to be fundamentally altered. If the international institutions currently involved in forest issues are unwilling or unable to take on this challenge, a new institution will have to be forged.

Chapter 12

The Depths of Europe

Lessons for North America

Human rights and environmental protection were among the most controversial issues discussed during the national debate over ratification of the North American Free Trade Agreement. Many of NAFTA's critics feared that the agreement would encourage NAFTA nations to attract and retain investment by lowering the cost of doing business. According to NAFTA's critics, this would result in downward harmonization or worker and environmental standards, and a race to the bottom. To lure investors, labor unions would be suppressed to keep wages down, worker safety standards would be reduced, and environmental protection requirements would be relaxed.

Although NAFTA was ratified by the U.S. Congress in December 1993, many of its critics predictions have proven accurate.[1] In 1996, the Washington D.C.-based Institute for Policy Studies (IPS) released a comprehensive report entitled *NAFTA's First Two Years*.[2] IPS reported that, since the agreement's adoption, there has been a demonstrable decline in environmental protection and workers' rights in all of NAFTA's countries, including the United States. In the United States, perhaps the most visible sign of this decline has been the recent Congressional effort to lower environmental standards to increase international competitiveness.

As Canada, the United States, and Mexico confront the human rights and environmental consequences of NAFTA, they might seek guidance from another regional trade regime—the European Union (EU). As an older trade regime, the EU has already encountered and attempted to resolve many of the problems currently facing NAFTA. The European experience teaches one basic lesson. To achieve just and sustainable trade policies, international regimes must possess the authority to implement human rights and environmental measures. Unfortunately, NAFTA lacks this authority.

The Limits of NAFTA

In its present form, NAFTA is primarily an agreement, not a political institution. It is a document that seeks to protect and promote the unregulated transnational trade of goods among Canada, the United States, and Mexico. It does so by restricting certain types of trade restrictions, such as subsidies, import tariffs, and quantitative import restrictions. The main institutions created to implement NAFTA, such as dispute resolution panels, possess primarily negative powers. These tribunals can determine that a NAFTA nation is in violation of the agreement, but they cannot adopt new international standards.

NAFTA's side agreements on labor and the environment suffer similar and even more pronounced institutional weaknesses. Under the North American Agreement on Labor Cooperation (NAALC) and the North American Agreement on Environmental Cooperation (NAAEC), international commissions are established to determine whether NAFTA nations are suppressing labor rights or lowering environmental standards to attract investment. Such actions are prohibited under the terms of the side agreements, as well as under provision in NAFTA.

The NAALC and NAAEC commissions, however, cannot adopt new international standards. Their power is limited to investigating alleged labor and environmental violations, and then seeking to bring violative nations into compliance. Unfortunately, the NAALC and NAAEC commissions are not provided with the means to even achieve these rather limited objectives. For instance, under the NAAEC, the process for imposing sanctions is cumbersome, most likely requiring years of cooperative negotiations before any penalties will be assessed. Moreover, even if these penalties are eventually imposed, they are limited to $20 million for the first year. In the context of trillion dollar national economies, penalties of this amount are without teeth.

In short, NAFTA and its side agreements are focused on one primary goal—to preserve unregulated international trade. NAFTA is not currently designed to move proactively, to create new binding standards that could bring human rights and environmental concerns into the regional economic planning process. The narrowness of the NAFTA regime has hindered the creation of a just and sustainable trade policy in North America. Recent developments in Canada and Mexico illustrate the regime's shortcomings.

In Canada, the temperate rainforests of British Columbia are being logged at an environmentally unsustainable pace, degrading watersheds and destroying critical habitat for numerous endangered species. In the North American environmental community, there is widespread recognition that this logging is in violation of B.C., federal Canadian and international environmental law.[3] British Columbia's unwillingness to effectively enforce its own environmental laws, which lowers the logging industry's business costs, is also placing pressure to lower forest protection standards in the United States.

Current forest practices in British Columbia would appear to constitute a violation of both NAFTA and the NAAEC. NAFTA states that it is inappropriate to "encourage investment by relaxing health, safety or environmental measures." NAAEC requires that each country "effectively enforce its environmental laws and regulations."

While these guarantees sound good on paper, there are no institutions to effectively implement these provisions. Under NAFTA and the NAAEC, the most environmentalists can do is request that a side agreement commission undertake an investigation.

In Mexico, the uprising in Chiapas illustrates the NAFTA regimes' impact on human rights. For several decades, indigenous groups in Mexico have been struggling to maintain their communities and traditional farming rights from encroachment by the Mexican national government.[4] In the 1980s, the Mexican government launched a program to modernize agriculture, and began reallocating formerly traditional farmland to large corporate agriculture and ranching operations. Many of these operations were financed by foreign companies and investors, particularly from the United States. The national government's power to strip local farmers of their land was greatly enhanced in 1992, when Article 27 of the Mexican Constitution was amended and weakened.[5] Prior to its amendment in 1992, Article 27 had provided the legal basis for local farmers' land rights. Divested of their land rights, many local and indigenous communities were displaced by large agribusiness operations. Indigenous efforts to organize themselves as legitimate labor, human rights and political groups were suppressed by the Mexican government.

On January 1, 1994, the same day that NAFTA went into effect, the Zapatista Army of National Liberation occupied several towns in Chiapas, Mexico.[6] When Mexican President Ernesto Zedillo launched a counterattack againt the Zapatistas in early 1995, the rebels retreated to the east, into the Lacandon jungle. Although the situation has not escalated to a state of full civil war, warfare between the Zapatistas and the Mexican national government continues to this day. The Zapatistas continue to challenge the government's authority, maintaining that they were wrongfully dispossessed of their land. The situation in Chiapas raises several difficult questions. Should U.S. business interests be a party to Mexico's continuing suppression of labor and human rights? Should there be new North American institutions or provisions to ensure that foreign investment does not lead to the extinction of indigenous culture? The NAFTA regime, with its focus on preserving free trade, is ill-equipped to ask, let alone answer, these questions.

Deep Integration in Europe

The European trade regime began with a limited economic mandate quite similar to that of NAFTA. In 1957, the Treaty of Rome created the European Community (EC) to help reduce trade barriers and encourage regional economic development. Unlike NAFTA, however, the Treaty of Rome created more than a list of prohibitions. It created four new multinational political institutions: the European Commission, the European Parliament, the European Council of Ministers, and the European Court of Justice. Collectively, these institutions possessed the power not only to determine violations, but to adopt new all-European standards, called "regulations" or "directives."

Although the EC institutions initially focused on regulating trade and competition, they soon expanded into other related areas. EC regulations and directives were adopted relating to the rights of workers to organize politically, the labeling of hazardous substances, air pollution from industrial plants and drinking water quality. The EC's authority to adopt these regional standards was based on two provisions in the Treaty of Rome: Article 100, which authorizes EC legislation that "directly affects the establishment or functioning of the common market," and Article 235, which authorizes actions "necessary for community objective."

In the area of labor and human rights, for instance, the European Community adopted the 1989 Social Charter, which expressly provides workers the rights of association to constitute professional organizations or trade union to defend their economic and social interests.[7] The Social Charter further provides that every worker shall have the freedom to join—or not to join—such an organization without the threat of any personal or occupational disadvantage. Similarly, in 1983 the EC adopted a directive that mandated that nations protect workers from health risks associated with exposure to industrial chemicals.[8]

Significant progress has also been made in the environmental field. In 1987, the EC adopted the Single European Act,[9] and in 1991 it adopted the Maastricht Treaty.[10] These agreements expanded the EC's law-making powers, and also changed the EC to the European Union (EU). Article 130 of the Single European Act establish several new objectives for the EU, including " to preserve, protect, and improve the quality of the environment" and to "contribute toward protecting human health." The Maastricht Treaty provided the EU with additional powers to "ensure a prudent and rational utilization of natural resources," and to "promote, at the international level, measures to deal with regional or worldwide environmental problems."

As a result of these provisions, Europe has adopted numerous pieces of environmental legislation, dealing with such issues as hazardous waste shipment and disposal, agricultural waste, urban noise, water pollution, environmental labeling and species habitat conservation. Moreover, in 1991, a European Environmental Agency (headquartered in Copenhagen, Denmark) was established to collect information on environmental protection, and to help monitor national compliance with EU directives and regulations. This new agency will compliment the existing policy efforts of the European Commission's Environmental Directorate (DG XI), which already plays an active role in developing and proposing EU legislation.

In addition to these integrative treaties, directives and institutions, the EU has also created an effective forum to enforce human rights and environmental guarantees—the European Court of Justice. Unlike the trade tribunals under NAFTA, or the commissions under NAFTA's side agreements, the ECJ has jurisdiction over all disputes arising under EU law. Moreover, the ECJ grants standing not only to national governments, but to private organizations and citizens that can establish that a challenged action is of direct and individual concern.[11] Although many believe that these standing requirements should be relaxed even further, ECJ rulings have nonetheless helped give teeth to the EU's human rights and environmental provisions.

For instance, in the 1987 case of *UNECTEF v. Heylens*, the ECJ ruled that unreasonable restrictions on a workers' right to relocate and change jobs violated EU law.[12] As another example, in the 1988 case of *European Commission v. Denmark*, the ECJ reviewed a Danish law requiring that nations desiring to export bottled beverages into Denmark had to have programs to recycle and reimport these same bottles once they had been used.[13] Although the law did interfere with the open flow of goods, the ECJ upheld the law as consistent with the EU's larger objectives. Similarly, in the 1990 case of *European Commission v. Belgium*, the ECJ considered a regional Belgian law banning the importation of wastes from other regions. In finding that the Belgian law was compatible with EU obligations, the ECJ held that "the accumulation of waste even before it reaches levels that will present dangers to health, constitutes a danger to the environment, especially when considering the limited capacity of each region or locality to receive them."[14]

By expanding the types of issues it may regulate, and by creating institutions that enable the creation of new human rights and environmental standards, the EU has evolved into something much more than a mere free trade agreement. From its initial inception in 1957 as a vehicle to promote unregulated transnational trade, the EU has matured into a comprehensive multinational institution. It now has the power and means to integrate human rights and environmental protection into Europe's larger development framework. As economist C. Ford Runge noted in his 1994 book *Freer Trade, Protected Environment*, "Despite a mixed record on both trade and the environment (in a world where no government is free of guilt), the EU has achieved a level of integration of both areas that merits careful attention. Its experience offers evidence to support the possibility of balancing the forces of trade integration and environmental protection."[15]

The key to Europe's integration process has been the recognition that the free movement of goods is merely one among many equally valid public policy goals. In this framework, issues such as environmental protection need not work around the principle of unrestricted trade. As Swiss legal scholar Andreas Ziegler has observed, although the ECJ initially legitimized environmental measures by referring to the possible trade effects of diverging national standards, it later considered the protection of the environment to be an essential Community objective which legitimized certain restrictions on the free movement of goods.[16] Within this framework, environmental protection is viewed as an appropriate and independent basis for regulating, rather than merely as an obstacle for infringing upon, international trade.

This political maturation process is often referred to as "deep integration".[17] Under deep integration, a region moves beyond the initial goal of removing trade obstacles, and looks to create an international framework to govern a broader range of activities, such as human rights and environmental protection.[18] Although some have argued that this deepening can lead to downward harmonization,[19] to a lowering of standards, the European example suggests that this need not be the case. If a progressive consensus and firm political will are present, deep integration can provide an effective means to actually curtail the destructive tendencies of international trade.

Rethinking North America

To deal with the human rights and environmental protection issues raised by transnational trade, Canada, the United States and Mexico need to develop an integrated, comprehensive framework. The foundation of this framework should not be an unyielding adherence to regional free trade. Rather, it should be the principle of just and sustainable economic development. Because of its narrow free trade focus and its institutional weaknesses, the NAFTA regime currently cannot provide this framework.

What is needed is a North American institution with the broad objectives and legislative powers of the European Union—perhaps a North American Union (NAU). In the context of an NAU, NAFTA would not be the regional constitution upon which all future efforts must comply. Rather, NAFTA would simply be one aspect, one legislative component, of the NAU's larger mandate. Under an NAU-type framework, treaties like the NAAEC and the NAALC would possess the same authority as NAFTA, and would not be treated as subordinate or side agreements. Moreover, the new regime would provide an effective forum for addressing other non-free trade issues, and for proposing new regional initiatives.

The creation of an NAU with broad powers cannot happen overnight. As the evolution of the European Union demonstrates, nations are understandably reluctant to delegate law-making authority to untested international institutions. This delegation or sharing of legal authority is often viewed as a threat to national sovereignty. However, if international institutions are responsive to the needs of citizens, and they lead to more just and sustainable policies, this national reluctance can be overcome. For North America, the first step is to move beyond the narrowness of NAFTA, and beyond the institutional weaknesses of the NAAEC and NAALC, and to lay the foundation for a more comprehensive and democratic regional regime.

Canada, the United States and Mexico should learn from Europe's experience. As the ECJ declared in the 1985 case of *Procureur de la Republique v. ADBHU*, when it upheld a French law that regulated the recycling of waste oil, "the principle of freedom of trade is not to be viewed in absolute terms but is subject to certain limits justified by the objectives of general interest pursued by the Community."[20] The creation of a new North American institution with a similar broad mandate is the best means to ensure that human rights and environmental protection are an integral part of the region's agenda.

Conclusion
Commonplace Ideas

In the environmental context, policy debates often come down to the issue of sovereignty. Property owners contend that they have a sovereign right to use their property in any manner they deem appropriate. Multinational corporations contend that they have a sovereign right to an unrestricted international marketplace. Nations claim that they have a sovereign right to determine how they manage their natural resources and environment. In brief, the issue of sovereignty is framed primarily in terms of rights.

Yet, there is another component to sovereignty—responsibility. At the individual level, citizens have rights but they also have responsibilities. They may not act in such a manner, or use their property, so that they damage the health or welfare of other citizens. Their entitlement to rights is contingent upon the recognition of their obligations to their neighbors and society. At the national corporate level, companies are not free to disregard free speech, labor or environmental laws in their pursuit of profit. As such, a corporation's right to develop economically is contingent on its willingness to honor these obligations and responsibilities. Although many would argue that our national policies fall far short of ensuring that these obligations are met, the nexus between rights and responsibilities is nonetheless an established part of our national political framework.

This notion of responsibility and obligation is not yet an established part of the international framework. Multinational corporations that conduct international business in a manner that deliberately seeks to increase environmental degradation and undermine human rights are nonetheless protected by trade rules. Their sovereign right to pursue profit is not contingent on any responsibility to the global community. Similarly, nations that accelerate global warming, deforestation, and species extinction, and that routinely deny human rights, are nonetheless welcome at the international table. Their citizenship among nations is not called into question.

The point here is not to demonize the market or corporations. The point is to recognize that, both conceptually and politically, there has been a failure at the international level. At this point in time, we have not forged a moral consensus that corporations and nations have fundamental responsibilities to the larger global community, to future generations, and to the other species with whom we share the earth. If this consensus existed, then the international institutions and rules to implement this consensus would be in place. If this consensus existed, then the debate could move beyond the safe rhetoric of sustainability to a frank assessment of international reforms.

In the legal field, some steps have been taken to help forge this consensus. For instance, at an International Law Conference held in the Netherlands in 1991, the participants adopted the *Hague Recommendations on International Environmental Law*. The *Hague Recommendations* stated: "It should be acknowledged as a rule that the principle of sovereignty implies the duty of a state to protect the environment within its jurisdiction, the duty to prevent transboundary harm, and the duty to preserve the global commons for present and future generations."

We can also point to Judge Weeramantry's opinion in the 1997 Danube River case before the International Court of Justice (ICJ). In reviewing whether Hungary was entitled to withdraw from an agreement to build a hydro-electric dam, Judge Weeramantry held that "the protection of the environment is a vital part of contemporary human rights doctrine, for it is the *sine qua non* for numerous human rights such as the right to health and the right to life itself."[1] Concluding that "environmental rights are human rights," he held that "there is a duty lying upon all members of the community to preserve the integrity and purity of the environment."

In this language of the *Hague Recommendations*[2] and of Judge Weeramantry's ICJ opinion, we can see the seeds of a broader concept of citizenship in the global community. If accepted, this concept would result in many changes. Economically developed nations would take responsibility for the role their current consumption and investment play in degrading the environment of the developing world. Multinational corporations would find their right to participate in the global marketplace conditioned on compliance with practices that preserve the environment and respect human rights. Nations that do not honor their international environmental and human rights obligations would find their citizenship among nations revoked, along with the trade benefits that accompany such citizenship.

Forging this moral consensus will not be easy. There are powerful interests hard at work to ensure that such an effort fails. The tide of change, however, is not with these interests. Ecology is revealing that environmental degradation has direct and profound international consequences, regardless of where such degradation initially takes place. Economics is providing us with greater understanding of the way international monetary policy, trade rules, and corporate investment contribute to pollution and resource abuse. In the diplomatic arena, there is a growing recognition that resource degradation and profound economic inequities are inherently destabilizing.[3]

As former U.S. Secretary of State Warren Christopher recognized in a 1996 speech at Stanford University, there is now an ecological dimension to national security. Christopher explained, "We must contend with the vast new danger posed to our national interest by damage to the environment and the resulting global and regional instability. . . . Our ability to advance our global interests is inextricably linked to how we manage the earth's natural resources. That is why we are determined to put environmental issues where they belong: in the mainstream of foreign policy."[4]

The modern lessons of ecology, economics and diplomacy are cumulative and reinforcing. They suggest that, although slowly, we are moving in the direction of enforceable environmental obligations at the international level. Notions of responsibility and obligation have begun to infuse our definitions of citizenship and sovereignty. It is imperative that this process be strengthened and accelerated, so that we can begin the process of developing responses equal to the scope and severity of our environmental problems.

Selected Resource Guide

Publications

The Amicus Journal
Natural Resource Defense Council
20 West 20th Street
New York, NY 10011
212-727-2700

Earth Island Journal
300 Broadway, Suite 28
San Francisco, CA 94133
415-788-3666

The Ecologist
Agricultural House
Bath Road
Sturminster Newton
Dorset, England
DT10 1DU
44-258-473-795

Ecology Law Quarterly
Boalt Hall School of Law
University of California
Berkeley, CA 94729
510-642-0457

The Environmental Forum
Environmental Law Institute
1616 P Street, Siute 200
Washington, DC 20036
202-328-5150

Environmental Policy and Law
c/o International Council on Environmental Law
Adenauerallee 214, D-5113 Bonn
Germany

Georgetown International Environmental Law Review
Georgetown University School of Law
600 New Jersey Avenue NW
Washington, DC 20001
202-662-9689

International Environmental Affairs
University Press of New England
23 South Main Street
Hanover, NH 03755-2048

Journal of Environmental Law
Oxford University Press
Walton Street
Oxford OX2 6DP
United Kingdom

Review of European Community & International Environmental Law (RECEIL)
Foundation for International Environmental Law & Development (FIELD)

SOAS/University of London
46-47 Russell Square
London WC1B4JP
United Kingdom
44-71-637-7950

Organizations

B.C. Wild
Box 2241, Main Post Office
Vancouver, B.C.
Canada, V6B 3W2
604-669-4802

California Environmental Trust
Hearst Building
5 Third Street, #608
San Francisco, CA 94103
415-543-1855

Canadian Institute for Resources Law (CIRL)
PF-B 3330
The University of Calgary
Calgary, Alberta T2N 1N4
Canada
403-220-3200
E-mail: cirl@acs.ucalgary.ca

Center for International Environmental Law (CIEL)
1621 Connecticut Avenue NW
Washington, DC 20009-1076

University of Saskatchewan
51 Campus Drive
Saskatoon SK s7n 5a8
Canada
306-96608893
E-mail:jck@fc.usask.ca

Communities for a Better Environment (CBE)
500 Howard Street, #504
San Francisco, CA 94105
415-243-8373
email: cbelegal@igc.apc.org

Earth Island Institute
300 Broadway, #28
San Francisco, CA 94133
415-788-3666

Earthjustice Legal Defense Fund
180 Montgomery Street, Suite 1400
San Francisco, CA 94104
415-627-6700

Environmental Defense Fund
257 Park Avenue South
New York, NY 10010
212-505-2100

Environmental Law Alliance Worldwide (E-LAW)
1877 Garden Avenue
Eugene, OR 97403
503-687-8454
email: elaw.usoffice@conf.igc.apc.org

Forest Stewardship Council
P.O. Box 849
Richmond, VT 05477
802-244-6257

Foundation for International Environmental Law & Development (FIELD)
SOAS/University of London
46-47 Russell Square
London WC1B4JP
United Kingdom
44-71-637-7950
email: field@gn.apc.org

Global Environmental & Trade Study (GETS)
Yale University
205 Prospect Street
New Haven, CT 06511
203-776-8167
email: scharnovitz@gets.org

Greenbelt Alliance
116 New Montgomery Street, #640
San Francisco, CA 94105
415-543-4291

International Forum on Globalization
950 Lombard Street
San Francisco, CA 94133
415-771-3394

International Rivers Network
1847 Berkeley Way
Berkeley, CA 94703
510-848-1155

International Union for the Conservation of Nature (IUCN)
Environmental Law Centre
Adenauerallee 214
D-5300 Bonn 1
Germany

National Parks & Conservation Association
1776 Massachusetts Avenue NW
Washington, DC 20036
202-223-6722
E-mail: natparks@aol.com

National Roundtable on the Environment and the Economy (NRTEE)
1 Nicholas Street Suite 1500
Ottawa, ON K1N7B7
Canada
613-992-7189

National Wildlife Federation
1400 16th Street NW
Washington, DC 20036-2266
202-797-5486
http://www.nwf.org

Natural Heritage Institute
114 Sansome, #1200
San Francisco, CA 94104
415-288-0550
email: nhi@igc.apc.org

Natural Resources Defense Council (NRDC)
40 West 20th Street
New York, NY 10011
212-727-2700
http://www.nrdc.org

Nautilus Institute
1831 Second Street
Berkeley, CA 94710
510-204-9296
http://www.nautilus.org

Northwest Ecosystem Alliance
P.O. Box 2813
Bellingham, WA 98227
206-447-1880
email: nwwatch@igc.apc.org

Northwest Environmental Watch
1402 Third Avenue, Suite 1127
Seattle, WA 98101-2118
206-447-1880

Pacific Environment and Resources Center (PERC)
1055 Fort Cronkhite
Sausalito, CA 94965
415-332-8200
perc@igc.apc.org

Rainforest Action Network
450 Sansome, Suite 700
San Francisco, CA 94111
415-398-4404
E-mail: rainforest@igc.apc.org

Resource Renewal Institute
Building A–Fort Mason Center
San Francisco, CA 94123
415-928-3774
http://www.rri.org

Sierra Legal Defense Fund
207 West Hastings Street, Suite 601
Vancouver, BC
Canada V6B 1H6

Transportation Alternatives
92 St. Marks Place
New York, NY 10009
212-629-3311

The Trust for Public Land
116 New Montgomery, 4th Floor
San Francisco, CA 94105
415-495-4014

Western Environmental Law Center
1216 Lincoln Street
Eugene, OR 97401
503-485-2471
email: westernlaw@igc.org

World Resources Institute
1709 New York Avenue NW
Washington, DC 20006
202-662-3499

Web Sites

Earthlaw, Environmental Law in the Public Interest
http://www.earthlaw.org
(information on private enforcement of U.S. environmental laws)

EcoNet
http://www.igc.org/igc/econet/index.html
(environmental news and feature articles)

Environmental News Link
http://www.caprep.com/caprep
(news update on legislation and court decisions)

Environmental Treaties and Resource Indicators
http://sedac.ciesin.org/pidb/pidb-home.html
(detailed information on international environmental treaties)

Multilateral Environmental Treaties
http://www.greenpeace.org/intlaw
(hypertext format of international environmental treaties, maintained by Greenpeace's legal department)

North American Commission for Environmental Cooperation (NACEC)
http://www.cec.org
(information on environmental law in Canada, Mexico and United States, as well as activities of the CEC)

United States Environmental Protection Agency
http://www.epa.gov
(information on all aspects of federal environmental law)

Selected Bibliography

Athansiou, Tom. *Divided Planet: The Ecology of Rich and Poor.* Boston: Little Brown, 1996.

Bredahl, Maaury (Ed.). *Agriculture, Trade & the Environment: Discovering and Measuring the Critical Linkages.* Boulder: Westview, 1996.

Cavanagh, John, Daphe Wysham, and Marco Arruda, *Beyond Bretton Woods: Alternatives to the Global Economic Order.* London: Pluto, 1994.

Daly, Herman. *Beyond Growth: The Economics of Sustainable Development.* Boston: Beacon, 1996.

Devall, Bill (Ed.). *Clearcut: The Tragedy of Industrial Forestry.* San Francisco: Sierra Club Books/Island Press, 1993.

Downs, Anthony. *New Visions for Metropolitan American.* Washington, D.C.: Brookings Institution Press, 1994.

Dudley, Nigel, Jean-Paul Jeanernaud, and Francis Sullivan. *Bad Harvest: The Timber Trade and the Degradation of the World's Forests.* London: Earthscan, 1995.

Esty, Daniel C. *Greening the GATT: Trade, Environment and the Future*. Washington, D.C.: Institute of International Economics, 1994.

George, Susan. *A Fate Worse Than Debt*. New York: Grove Weiderfeld, 1990.

Humphreys, David. *Forest Politics*. London: Earthscan, 1996.

Johnson, Pierre Marc, and Andre Beaulieu. *The Environment and NAFTA* (Washington, D.C.: Island Press, 1996).

Maser, Chris. *The Redesigned Forest* (San Pedro: R. and E. Miles, 1987).

Pearce, David W., and R. Kerry Turner. *Economics of Natural Resources and the Environment*. Baltimore: Johns Hopkins University Press, 1990.

Porter, Gareth, and Janet Welsh Brown. *Global Environmental Politics*. Boulder, CO: Westview, 1996.

Petersen, D.J. *Troubled Lands: The Legacy of Soviet Environmental Destruction*. Santa Monica: RAND Publications, 1993.

Ross, Monique, and Owen Saunders. *Environmental Protection: Its Implications for the Canadian Forest Sector*. Calgary: Canadian Insitute of Resources Law, 1993.

Runge, C. Ford. *Freer Trade, Protected Environment*. New York: Council on Foreign Relations Press, 1994.

Simon, Joel. *Endangered Mexico: An Environment on the Edge*. San Francisco: Sierra Club Books, 1997.

Weiss, Edith Brown. *In Fairness to Future Generations: International Law, Common Patrimony and Intergenerational Equity*. New York: United Nations Press, 1989.

Yaffee, Stephen. *The Wisdom of the Spotted Owl: Policy Lessons for the Next Century*. Washington, D.C.: Island, Press, 1995.

Ziegler, Andreas. *Trade and Environmental Law in the European Community*. London: Oxford University Press, 1996.

Suggested Further Reading

Adams, Patricia. *Odious Debts: Loose Lending, Corruption, and the Third World's Environmental Legacy*. London: Earthscan, 1991.

Anderson, Terry L., and Donald R. Leal. *Free Market Environmentalism*. Boulder: Westview, 1991.

Attali, Jacques. *Millenium: Winners and Losers in the Coming World Order*. New York: Random House, 1991.

Bean, Michael. *The Evolution of National Wildlife Law*. New York: Praeger Publishers, 1983.

Bell, Daniel. *The Cultural Contradictions of Capitalism*. Cambridge: Harvard, 1976.

Berry, Wendall. *The Unsettling of America: Culture & Agriculture*. San Francisco: Sierra Club Books, 1977.

Bouchuan, He. *China on the Edge: The Crisis of Ecology and Development*. San Francisco: China Books, 1991.

Brenton, Tony. *The Greening of Machiavelli: The Evolution of International Environmental Politics*. London: Earthscan 1994.

Burbank, James. *Vanishing Lobo*. Johnson Books, 1990.

Caldwell, Lynton Keith. *International Environmental Poliocy: From the Twentieth to the Twenty-First Century*. Durham: Duke University Press, 1996.

Dauvergne, Peter. *Shadows in the Forest: Japan and the Political Economy of Deforestation in Southeast Asia*. Cambridge: MIT Press, 1997.

de Klemm, Cyrille, and Clare Shine. *Biological Diversity Conservation and the Law: Legal Mechanisms for Conserving Species and Ecosystems*. Bonn: IUCN Environmental Law Centre, 1995.

Dowie, Mark. *Losing Ground: American Environmentalism at the Close of the Twentieth Century*. Cambridge: MIT Press, 1994.

Friedman, Mitch, and Paul Lindholdt (Eds.). *Cascadia Wild: Protecting an International Ecosystem*. Bellingham, WA: Greater Ecosystem Alliance, 1993.

Goodman, David, and Michael J. Watts. *Globalizing Food: Agrarian Questions and Global Restructuring*. New York: Routledge, 1997.

Haas, Peter M., Robert O. Keohane, and Marc A Levy (Eds.). *Institutions for the Earth: Sources of Effective International Environmental Protection*. Cambridge: MIT Press, 1993.

Hanson, Victor Davis. *Fields Without Dreams: Defending the Agrarian Idea*. New York: Free Press, 1996.

Harrison, Robert Pogue. *Forests: The Shadow of Civilization*. Chicago: Univeristy of Chicago Press, 1992.

Irwin, Douglas. *Against the Tide: An Intellectual History of Free Trade*. Princeton: Princeton University Press, 1996.

Kay, Jane. *Asphalt Nation*. San Francisco: Sierra Club Books, 1996.

Kiss, Alexander, and Dinah Shelton. *Manual of European Environmental Law*. New York: Cambridge University Press, 1993.

Loomis, R., and M. Wilkinson. *Wildwood: A Forest for the Future*. Gabriola, British Columbia: Reflections, 1990.

Manser, Roger. *Failed Transitions: The Eastern European Economy and the Environment Since the Fall of Communism*. New York: New Press, 1994.

Matthews, Jessica Tuchman (ed.). *Preserving the Global Environment: The Challenge of Shared Leadership*. New York: W.W. Norton, 1991.

McLaughlin, Andrew. *Regarding Nature: Industrialism and Deep Ecology*. Albany: SUNY Press, 1993.

Moe, Richard, and Carter Wilkie. *Changing Places: Rebuilding*

Community in the Age of Sprawl. New York, Henry Holt, 1997.

Munoz, Heraldo, and Robin Rosenberg (eds.). *Difficult Liason: Trade and the Environment in the Americas.* New Brunswick: Transaction Books, 1993.

Nash, Roderick. *Wilderness and the American Mind.* New Haven: Yale University Press, 1982.

Organization for Economic Cooperation and Developmemnt (OECD). *The Environmental Effects of Trade.* Paris: OECD Publications, 1994.

Panaytou, Thomas, and Peter Ashton. *Not by Timber Alone.* Washington, D.C.: Island Press, 1992.

Power, Thomas. *Lost Landscapes and Failed Economies: The Search for a Value of Place.* Washington, D.C.: Island Press, 1996.

Reed, Peter, and David Rothenberg (eds.). *Wisdom in the Open Air: The Norwegian Roots of Deep Ecology.* Minneapolis: Univeristy of Minnesota Press, 1993.

Repetto, Robert. *Public Policy and the Misuse of Forest Resources* Washington, D.C.: World Resources Institute, 1988.

Rich, Bruce. *Mortgaging the Earth: The World Bank, Environmental Impoverishment and the Crisis of Development.* Boston: Beacon Press, 1994.

Rusk, David. *Cities Without Suburbs.* Washington, D.C.: Woodrow Wilson Center Press, 1993.

Russell, Sharman Apt. *Kill the Cowboy: A Battle of Mythology in the New West.* New York: Addison Wesley, 1993.

Sax, Joseph. *Defending the Environment.* New York: Alfred Knopf, 1971.

Swanson, Timothy M., and Edward B. Barbier. *Economics for the Wilds: Wildlife, Diversity and Development.* Washington, D.C.: Island Press, 1992.

Wilderness Society. *Federal Forests and the Economic Base of the Pacific Northwest: A Study of Regional Transition.* Seattle: Wildnerness Society Publications, 1992.

Wright, R. Gerald (ed.). *National Parks and Protected Areas: Their Role in Environmental Protection*. Cambridge: Blackwell Science, 1996.

Zaelke, Durwood, Paul Orbuch, and Robert Housman (Eds.). *Trade and the Environment: Law, Economics and Policy*. Washington, D.C.: Island Press, 1993.

Notes

Introduction:
Sharp Teeth

1. *Crossing the Next Meridian: Land, Water and the Future of the West;* and *American Indians, Time and the Law.*
2. Des Pres. *Writing Into the World: Essays 1973–1987*. Viking Penguin, 1991.

Chapter 1:
City Limits: Urban Ecology and Economic Justice

1. Downs. *New Visions for Metropolitan America.* 1996, p. 47.
2. *Beyond Sprawl: New Patterns of Growth to Fit the New California.* 1995 Report by Bank of America, The California Resources Agency, Greenbelt Alliance and the Low Income Housing Fund, hereinafter *Beyond Sprawl*, pp. 1–5.
3. Hayward. *Preserving the American Dream: The Facts About Suburban Communities and Housing Choice.* September 1996 Report by the California Building Industry Association, hereinafter *CBIA Report*, p. 1.
4. Rybczynski. *City Life*. 1995, p. 176.
5. Rybczynski. *City Life*. 1995, pp. 190-197.
6. Mumford. *The City in History: Its Origins, Its Transformation and Its Prospects.* 1961, p. 486.

7. *Id*. at 496.
8. Davis. *City of Quartz: Excavating the Future in Los Angeles*. 1992, p. 131.
9. *Beyond Sprawl: New Patterns of Growth to Fit the New California*. 1995 Report published by Bank of America, the California Resources Agency, Greenbelt Alliance and the Loaw Income Housing Fund, p. 2.
10. *Four Reasons Why Bay Area Business Should Push for Regional Growth Management*. 1991 publication by the Bay Area Council, hereinafter Bay Area Council Report.
11. *Id*.
12. Tarlock. "City Versus Countryside: Environmental Equity in Context," 21 *Fordham Urban Law Journal* 461. 1994.
13. Davis. *City of Quartz: Excavating the Future in Los Angeles*. 1992, p. 173.
14. 42 U.S.C.A. Sections 9601-9675 (1988).
15. "Superfund as a Threat." *The Environmental Forum*. (July/August 1994), p. 15.
16. The term "brownfields" is derived from the rusting building and machinery and petroleum-contaminated property generally associated with abandoned industrial sites. The term is often used to distinguish between property previously used for industrial purposes and undeveloped pristine property (often called "greenfields").
17. 42 U.S.C.A. 9601(35)(A)(West Supp. 1992).
18. Smith. "CERCLA's Innocent Landowner Defense: Oasis or Mirage?" 18 *Columbia Journal of Environmental Law* 155. 1993.
19. The Brownfields Action Agenda. April 1996 Document by the U.S. Environmental Protection Agency, p. 1.
20. The Brownfields Action Agenda. April 1996 Document by the U.S. Environmental Protection Agency.
21. EPA tracks potentially contaminated sites on CERCLIS (Comprehensive Environmental Response, Compensation and Liability Information System).

22. "Bringing New Life to a Troubled Area." *The San Francisco Chronicle*. Aug. 29, 1994, p. A18.
23. Notes from Author's Interview with Alan Edson and Olin Webb, Feb. 10, 1997.
24. Downs. *New Visions for Metropolitan America*. 1996, p. 31.
25. Author's interview with Joe Bodovitz. March 7, 1997.
26. 15 U.S.C. Section 637(a).
27. *Key Lessons for Brownfields From the Base Closure Cleanup Process*. Fact Sheet Prepared by Lenny Siegel of the CAREER/PRO, a program of the San Francisco Urban Institute.
28. *Restoration Advisory Board Implementation Guidelines*. Department of Defense and Environmental Protection Agency. September 1994. See also "Naval Center Seeks Help From Public." *The Indianapolis Star*. March 5, 1996, p. B3.

Chapter 2:

Roughshod: Northwest Forests and the Constitution

1. Section 2001, 1995 Recissions Act.
2. 80 U.S. 128 (1871).
3. 1990 Department of the Interior and Related Agencies Appropriations Act. Section 318(b)(6)(A).
4. 914 F.2d 1311 (1990).
5. 914. F.2d at 1315 (9th Cir. 1990).
6. 112 S.Ct. 1413 (1992).
7. 112 S.Ct. at 1409 (1992).
8. Ronner. "Judicidial Self-Demise: The Test of When Congress Impermissibly Intrudes on Judicial Power After Robertson v. Seattle Audubon Society." *Arizona Law Review*. 1993.
9. Rutzick, "Salvaging the Logging Rider." *Legal Times*. July 15, 1996, p. 24.
10. *Submission Pursuant to Article 14 of the North American Agreement on Environmental Cooperation on the United*

States' Logging Rider. Prepared by Sierra Club Legal Defense Fund, August 1995.

11. Stahmer. "Clinton Gives In to Republican Pressure and Accepts Salvage Rider, But Directs Forest Service and BLM to Obey All Environmental Laws and Forest Plans Anyway," *Headwaters Journal.* Fall 1995, p. 3-7.

Chapter 3:

Blaming Wildlife: The Endangered Endangered Species Act

1. Hueber. "American Report High Levels of Environmental Concern." Activity. *The Gallup Poll Monthly,* No. 307. April 1991, p. 6.
2. There were several regional environmental groups that opposed Option 9, and thus continued to litigate environmental and forest protection issues surrounding logging of forests in the Pacific Northwest. These groups included the Native Forest Council, the Kalmiosis Audobon Society, and the Friends of the Brietenbush Cascades. *See* Tokar, "Between the Loggers and the Owls: The Clinton Northwest Forest Plan." *The Ecologist.* July/August 1994, p. 149.
3. Dinwoodie. "The Endangered Species Act—A Law That Works." *Headwaters Journal.* Winter 1994–95, p. 30.
4. Anderson and Olson. *Federal Forests and the Economic Base of the Pacific Northwest: A Study of Regional Transitions.* The Wilderness Society, 1992.
5. Dinwoodie. "The Endangered Species Act—A Law That Works." *Headwaters Journal.* Winter 1994–95, p. 30.
6. Yaffee. *The Wisdom of the Spotted Owl: Policy Lessons for the Next Century.* Island Press, 1994, p. 160.
7. Pissot. "Timber Troubles: The Spotted Owl is Not the Cause of the Northwest Forest Crisis." *The Washington Post.* April 2, 1993.
8. Irvin. "The Endangered Species Act: Keeping Every Cog and

Wheel." *Natural Resources and Environment*. American Bar Association, Summer 1993, p. 40.

9. Wilson. *Biodiversity*. Washington, D.C: National Academy of Sciences Press, 1986.
10. Desiderio. "The ESA: Facing Hard Truths and Advocating Responsible Reform," *Natural Resources and the Environment*. Summer 1993, p. 41.
11. Kubasek, Browne, and Mohn-Klee. "The Endangered Species Act: Time for a New Approach?" *Environmental Law*. 1994, p. 337.
12. Mueller. "Regulatory Abdication." *The San Francisco Recorder*. November 1997 Environmental Law Magazine, p. 7
13. Pfaff. "Landmark Changes." *The Daily Journal*. October 1997, Environmental Law Supplement, p. 35.

Chapter 4:

Words to Choke On: Free Speech and Environmental Debate

1. *National Advertisers v. State of California*, 809 F. Supp. 747 (1992).
2. California Business & Professions Code, Section 17508.5
3. 44 F.3d 726 (9th Cir. 1994)
4. 44 F.3d at 731 (9th Cir. 1994).
5. Bleifuss. "Will Ronald Eat McCrow?" *In These Times*. Dec. 25, 1994, p. 17.
6. Vick and Macpherson. "An Opportunity Lost: The United Kingdom's Failed Reform of Defamation Law." 49 *Federal Communications Law Journal* 621. April 1997. See also *Mapp v. News Groups Newspapers Limited*. Court of Appeal, Civil Division, March 1997.
7. Zoll. "Big Mac Attack: A British Libel Trial Puts McDonald's on the Grill." *The San Francisco Bay Guardian*. January 29, 1997, p. 22.

8. Graffy. "Big Mac Bites Back: English Law and an Experienced Lawyer Help Burder Giant Win Case Againt Small-Fry Environmentalists." *American Bar Association Journal.* August 1997, p. 22.
9. Vick & Macpherson. "An Opportunity Lost: The United Kingdom's Failed Reform of Defamation Law." 49 *Federal Communications Law Journal* at 628. April 1997.

Chapter 5:

Ignorance Abroad: International Projects Under National Law

1. Ernsdoff. "The Agency for International Development and NEPA: A Duty Unfulfilled." 67 *Washington Law Review* 133. 1992.
2. *Sierra Club v. Adams.* 587 F.2d 398 (D.C. Cir. 1978); *Wilderness Society v. Morton*, 436 F.2d 1261 (D.C. Cir. 1972); *Environmental Defense Fund v. Masey*, 986 F.2d 528 (D.C. Cir. 1993).
3. *Greenpeace USA v. Stone*, 748 F. Supp. 749 (D. Hawaii 1990); *Natural Resources Defense Council v. Nuclear Regulatory Commission*, 647 F.2d 1345 (D.C. Cir. 1981).
4. NEPA, 102.
5. Gordon. "U.S. Venture Must Protect Russian Taiga." *Oregonian.* Jan. 18, 1995, at A8.
6. Riechel. "Government Hypocrisy and the Extraterritorial Application of NEPA," 26 *Case Western Reserve Journal of International Law.* 115 1994, p. 118.

Chapter 6:

Axe to the Myth: Canadian Logging and International Law

1. McCrory. "Canada—Brazil of the North." *Clearcut: The Tragedy of Industrial Forestry.* 1994.

2. Cooperman. "Cutting Down Canada." *Clearcut: The Tragedy of Industrial Forestry*. 1994, p. 5.
3. *The State of Canada's Forests 1993*. Report by the Canadian Forest Service, p. 31.
4. *Clearcutting: A Canadian Perspective*. 1995 Report by Canadian Pulp and Paper Information Centre, Europe.
5. "Clearcutting Not As Bad as Critics Claiming." *Victoria Times-Colonist*. July 23, 1994, p. A5.
6. Noss. "Sustainable Forestry or Sustainable Forests?" *Defining Sustainable Forestry*. 1993 pp.17–18.
7. Gordon. "Ecosystem Management: An Idiosyncratic Overview." *Defining Sustainable Forestry*. 1993, pp. 240–42.
8. Harding. "Threats to Diversity of Forest Ecosystems in British Columbia." *Biodiversity in British Columbia: Our Changing Environment*. 1994, p. 257.
9. Biodiversity Convention, Article 8(c).
10. Biodiversity Convention, Article 8(d).
11. Biodiversity Convention, Article 9(b).
12. U.N. Statement of Forest Principles, Article 4.
13. U.N. Statement of Forest Principles, Article 2(b).
14. U.N. Statement of Forest Principles, Article 3(a).
15. Migratory Birds Convention, Sections 6(a) and (b).
16. Legal Memorandum to the Natural Resources Defense Council from the Yale Environmental Policy Clinic (February 22, 1995).
17. Twitchell. "Implementing the U.S.–Canada Pacific Salmon Treaty: The Struggle to Move from Fish Wars to Cooperative Fishery Management." 20 *Ocean Development and International Law* 420. 1989.
18. *B.C. Salmon Habitat: Conservation Plan*. September 1995 Strategy Paper published by the Province of British Columbia, pp. 2, 7.
19. NAAEC, Article 5; NAFTA Article 114.
20. 1985 Federal Fisheries Act; 1994 B.C. Forest Practice Code; 1993 Alberta Forest Conservation Strategy.

21. *B.C. Forestry Fact Sheet.* Sierra Club of Western Canada, 1993.
22. Dudley, Jeanrenaud and Sullivan. *Bad Harvest: The Timber Trade and the Degradation of the World's Forests.* Earthscan 1994, p. 64.
23. *Clearcut Sound: The Impact of Logging on the Environment and Wildlife.* Greenpeace International, 1994. p. 3.
24. Connely. "The Big Cut." *Sierra Magazine.* 1991, pp. 42–53.
25. Munro. "Report Finds B.C. Life Under Assault." *Toronto Star.* April 16, 1994, p. B6.
26. Dinnston. "The Temperate Rainforest: Canada's Clearcut Secret." *World Watch.* July/August 1993, p. 43.
27. Genovali. "Beyond Clayoquot Sound." *Earth Island Journal.* Summer 1995, p. 24.
28. Spalding. *Trade and the Environment: The British Columbian Timber Trade Example.* December 1994 Report by the Natural Resources Defense Council.
29. Nelson. "Taxpayers Stunningly Generous to Forest Firms." *Victoria Times Colonist.* October 13, 1993, p. A5.
30. Nelson. "Pulp and Propaganda." *The Canadian Forum.* July/August 1994, p. 14.
31. Drushka, Nixon, andTravers. *Touchwood: B.C. Forest at the Crossroads.* 1993.
32. *The State of Canada's Forest 1993.* Canadian Forest Service, p. 8.
33. Lederman. *The Courts and the Canadian Constitution.* 1964, p. 192.
34. Porter. *The Canadian Environmental Protection Act: Analysis of the Constitutional Deadlock Created by the Equivalency Provisions.* Unpublished 1992 report.
35. NAAEC, Annex 41.
36. Northey. "Federalism and Comprehensive Environmental Reform: Seeing Beyond the Murky Medium." 29 *Osgoode Hall Law Journal* 127. 1991.
37. Ross and Saunders. *Environmental Protection: Its Implications for the Canadian Forestry Sector.* CIRL, 1993, pp. 10.

38. Bernstein and Cashore. *The Internationalization of Domestic Policy Making: The Case of Eco-Forestry in British Columbia.* Paper prepared for presentation at the Annual Meeting of the Canadian Political Science Association, Brock University, St. Catharaines, Ontario, June 2–4, 1996.

Chapter 7:
Ecology After the USSR: Hard Times for Russian Environmental Law

1. Robinson. "Soviet Environmental Law and Perestroika," *Environmental Policy & Law*. 1988, p. 224.
2. The term "Siberia" is used here to indicate the entire region of the Russian Republic that lies east of the Ural Mountains. This defintion, although convenient for this article's present purposes, is not geographically precise. The Pacific Coast region of Russia is generally referred to as the Russian Far East, and is usually considered geographically separate from Siberia (which lies inland and further west).
3. Zimbler. "Legal Remedies Address a Catastrophic Situation: The Russian Law on Protection of the Environment." *CIS Environmental Watch*. Fall 1992, p. 41.
4. Levin. "Russian Forest Laws—Scant Protection During Troubled Times." 19 *Ecology Law Quarterly* 685. 1992.
5. Petersen. *Troubled Lands: The Legacy of Soviet Environmental Destruction.* 1993.
6. Gooding. "Bratsk Finds Going Tough in Real World—A Town That Depends Almost Entirely On its Outdated Smelter." *Financial Times*. Oct. 21, 1993, p. 34.
7. Block. "Russian Law: Does It Matter in the Wild East?." *Russian Far East Update*. January 1994, p. 7.
8. Zaharchenko. "The Environmental Movement and Ecological Law in the Soviet Union: The Process of Transformation," 17 *Ecology Law Quarterly* 455. 1990.

9. Deghan. "A Criticism of the New Mechanics for Environmental Protection in the Russian Federation." 19 *Review of Central and East European Law* 661. 1993.
10. Floroff and Teifenbrun. "Land Ownership in the Russian Federation: Laws and Obstacles." 37 *St. Louis University Law Journal* 235 (1993).
11. Grigoriev. "Russia's New Forestry Act: Leaving the Door Wide Open for Ruthless Exploitation." *Taiga News*. March 1993, p. 2.

Chapter 8:

United by Poison: Relief for Bhopal's Victims

1. Abraham and Abraham. "The Bhopal Case and the Development of Environmental Law in India." 40 *International & Comparative Law Quarterly* 334. 1991.
2. *Mass Disasters and Multinational Liability: The Bhopal Case.* Indian Law Institute, 1986.
3. Jacob. "Bhopal Gas Victims: Red Tape and Corruption Delay Compensation." *Down to Earth.* December 15, 1992, pp.5-6.
4. Rosencranz, Divan and Scott. "Legal and Political Repercussions in India." *Learning from Disaster: Risk Management After Bhopal,* University of Pennsylvania Press, 1994.
5. Cohen. "Nightmares and Hope in Bhopal." *Earth Island Journal.* Fall 1997, p. 23.

Chapter 9:

Refoliating Vietnam: A Second War for the Forests

1. Phan Nguyen Hong. *Causes and Effects of the Deterioration in the Mangrove Resource and Environment in Vietnam.* October 1994 Report, p. 24.
2. Belcher and Gennino. *Southeast Asia Rainforests: A Resource Guide and Directory,* Rainforest Action Network, 1993, p. 45.

3. Hiebert. "Taking Cover: Forestry Plays Small Role in Vietnam's Development." *Far Eastern Economic Review.* June 7, 1990, p. 46.
4. "UN Rings Alarm Bells on Vietnam Environment." *Reuters Limited.* July 4, 1994.
5. Sidel. "Law Reform in Vietnam: The Complex Transition from Socialism and Soviet Models in Legal Scholarship and Training." *11 U.C.L.A. Pacific Basin Law Journal* 221. 1993.
6. Gillespie. "Private Commercial Rights in Vietnam: A Comparative Analysis." 30 *Stanford Journal of International Law* 325. 1994.
7. *Biodiversity Protection in Vietnam.* Unpublished 1994 Report by World Wildlife Fund, p. 22.
8. Haeder "Vietnam's New War: Biodiversity Has a Small Claw-hold in Indochina, 20 Years After the Americans Left." *E Magazine.* September/October 1995, pp. 18-19.
9. Hiebert. "Food of Forests? Vietnamese Scientists Fear Shrimp-Raising Project." *Far Eastern Economic Review.* April 7, 1994.
10. Ly Tho. *Silicultural Characteristics of the Swamp Mangroves at Nam Can 0 Minh Hai Province.* October 1994 Report.
11. Jones. "Mekong Ecology Threatened By Siltation and Deforestation." *The Christian Science Monitor.* January 12, 1994, p. 12.
12. *Biodiversity Protection in Vietnam.* Unpublished 1994 Report by World Wildlife Fund, p. 43.
13. *Research Documents on Forest Inventory and Planning.* Vietnam Ministry of Forests and the Forest Inventory and Planning Institute, Hanoi 1991, pp. 89-90.

Chapter 10:
A Difficult Swim: The Sea Turtle Navigates GATT

1. *Earth Island v. Christopher,* 1996 Ct. Int'l Trade, LEXIS 71, SLIP. OP. 96-42. April 10, 1996.
2. "Turtles in the Soup." *The Economist.* March 16, 1996, p. 64.

3. Public Law No. 101-162, Title VI, 609, 103 Stat. 1988, 1037-38 (1989), codified at U.S.C.A. 1537 (1996 Supp.).
4. Batycki. "Trade War Over Turtles?." *Earth Island Journal.* Summer 1996, p. 9.
5. Irwin and Iundicello. *Delay and Denial: A Political History of Sea Turtles and Shrimp Fishing.* 1995.
6. "State Department Guidelines Implement Sea Turtle Decision." *BNA International Trade Reporter.* April 24, 1996, p. 687.
7. *The Kemp's Ridley Sea Turtle.* June 1996 Status Report by the Sea Turtle Expert Working Group, National Marine Fisheries Service.
8. "Ruling Seen Barring Most Shrimp Imports into U.S." *Reuters.* May 3, 1996.
9. 58 Fed. Re. 9,015-16 (Feb. 9, 1993).
10. Department of State. *Revised Notice for Guidelines for Determining Comparability of Foreign Programs for the Protection of Turtles in Shrimp Trawl Operations*, 61 Fed. Reg. 17,342. April 19, 1996.
11. *Earth Island v. Christopher*, Slip Opinion 96-165, Ct. of Int'l Trade, October 8, 1996.
12. *Canada Measures Affecting Exports of Unprocessed Herring and Salmon.* Report of the Panel adopted on 22 March 1988, GATT B.I.S.D. 35th Supp. (Geneva 1989)
13. *United States Restrictions on Imports of Tuna: Report of the GATT Panel.* Aug. 16, 1991, reprinted in 30 I.L.M. 1594.
14. Foy. "Toward Extension of GATT Standards Code to Production Processes." 26 *Journal of World Trade* 121. 1992; Esty. *Greening the GATT: Trade, Environment and the Future.* 1994, p. 51.
15. Charnovitz. "Green Roots, Bad Pruning: GATT Rules and Their Application to Environmental Trade Measures," 7 *Tulane Environmental Law Journal* 299. 1994.
16. Charnovtiz. "Environmental Harmonization and Trade

Policy." *Trade and the Environment: Law, Economics and Policy*. 1993, pp. 269-270.
17. Corben. "Thailand Targets, U.S. Curbs on Shrimp." *Journal of Commerce*. July 19, 1996.
18. "U.S. Wild Shrimp Embargo Looming." *Food Institute Report*. Feb. 19, 1996.

Chapter 11:

Trees Falling: Forests and the Timber Trade

1. Belecher and Gennino. *Southeast Asia Rainforests: A Resources Guide and Directory 1*. Rainforest Action Network and World Rainforest Movement, 1993, p. 1; *Forests in Trouble: A Review of the Statuts of Temperate Forests Worldwide*. World Wildlife Fund, 1992.
2. Esty. *Greening the GATT: Trade, Environment and the Future*. 1994, p. 18.
3. Rosencranz and Scott. "Siberia's Threatened Forests." *Nature*. January 1992, at 293.
4. Gordon. "U.S. Venture Must Protect Russian Taiga." *Oregonian*. January 18, 1996, p. A8.
5. Cushman. "Logging in Siberia Sets Off a Battle in the U.S." *The New York Times*. January 30, 1996, p. 3.
6. Collins and Lear. *Chile's Free Market Miracle: A Second Look*. 1995, pp. 211–213.
7. Warn, "Survey of Chile", *The Financial Times* (May 19, 1993) p. 41.
8. "Indonesia Welcomes Autrian Plan to Revoke Timber Law." *Reuters News Service*. March 2, 1993.
9. Hartwell and Bergkamp. "Eco-Labeling in Europe: New Market-Related Environmental Risks?" *International Environmental Reporter*. September 23, 1992, p. 623.
10. "Timber Industry: Euro Eco-Label Called a Trade Barrier." *Greenwire News Service*. June 23, 1995.

11. Martens. "Ending Tropical Deforestation: What is the Proper Role for the World Bank?" *Harvard Environmental Law Review* 485. 1989.
12. *Mainstreaming the Environment: The World Bank Group Since the Rio Earth Summit.* World Bank Publication, 1995.
13. Rich. "The Cuckoo in the Nest: Fifty Years of Political Meddling by the World Bank." *The Ecologist.* January/February 1994, pp.8–9.
14. Nectoux and Kuroda. *Timber From the South Seas: An Analysis of Japan's Tropical Timber Trade and its Environmental Impact.* 1989 World Wildlife Fund Report.
15. Browne. "Alternatives to the International Monetary Fund." *Beyond Bretton Woods.* 1994, pp. 57–73.
16. Humphreys. *Forest Politics: The Evolution of International Cooperation.* 1996, p. 133.

Chapter 12:

The Depths of Europe: Lessons for North America

1. Although NAFTA was never formally ratified by the Senate, as apparently required under the Constitution, it did receive an endoresement by a majority of the House of Representatives. This procedure is now a common method for approving international agreements. For a discussion of this change in the treaty approval process, see Ackerman and Golove. "Is NAFTA Constitutional?" 108 *Harvard Law Review* 801. 1995.
2. *NAFTA's First Two Years: The Myths and the Realities.* 1996 Report by the Institute for Policy Studies, pp. 11–33.
3. *Forests on the Line: A Comparison of Logging Practices in British Columbia and Washington State.* January 1995 Report by Vancouver's Sierra Legal Defense Fund and the Natural Resources Defense Council.
4. Ross. *Rebellion From the Roots: Indian Uprising in Chiapas.* 1995.

5. Simon. *Endangered Mexico: An Environment on the Edge.* 1997, p. 104.
6. "Mexico: Free Markets Versus Freedom." *Multinational Monitor.* April 1995, p. 28.
7. "Workers Rights in the EC Single Market." *The External Impact of European Unification.* January 1990, Buraff Publications, p. 13.
8. EC Directive on the Protection of Workers Against Risk of Exposure to Asbestos. 1983.
9. Kramer. "The Single European Act and Environmental Protection: Reflections on Several New Provisions in Communty Law." 24 *Common Market Law Review* 659. 1987.
10. Friedberg. "Closing the Gap Between Word and Deed in European Community Environmental Policy." 15 *Loyola of Los Angeles International & Comparative Law Journal* 2, 1993.
11. Lininger. "Liberalizing Standing for Environmental Plaintiffs in the European Union." 4 *New York University Environmental Law Journal* 96–101. 1995.
12. Case 222/86, 1987, ECR 4098.
13. Case 302/86, 1988 ECR 4067.
14. C-2/90. ECR I-443.
15. Runge. *Freer Trade, Protected Environment: Balancing Trade Liberalization and Environmental Interests.* 1994, p. 35.
16. Ziegler. *Trade and Environmental Law in the European Community.* 1996, p. 138.
17. Haas. "Why Collaborate? Issue Linkage and International Regimes," 32 *World Politics* 357. April 1980.
18. Reinicke. *Deepening the Atlantic: Toward New Transatlantic Marketplace?* July 1996, Brookings Institutions.
19. *The Case Against Free Trade: GATT, NAFTA and the Globalization of Corporate Power.* Earth Island Press, 1993.
20. Case 302/86 1988, ECR 4067.

Conclusion:
Commonplace Ideas

1. International Court of Justice: Judgement in Case Concerning the Gabcikovo-Nagymaros Project (Sept. 25, 1997), 37 International Legal Materials 162 (1998), p. 206.
2. The Hague Recommendations on International Environmental Law, Peace Palace, The Hague, August 16, 1991, Preamble.
3. Brown (Ed.). *In the U.S. Interest: Resources, Growth and Security in the Developing World.* 1990.
4. "American Diplomacy and the Global Environmental Challenges of the 21st Century," April 9, 1996 Address by Secretary of State Warren Christopher to Stanford University (Text Transcript, U.S. Department of State, Office of the Spokesman).

Index